Skills Interv
for Middle School Mathematics
Diagnosis and Remediation

Student Workbook

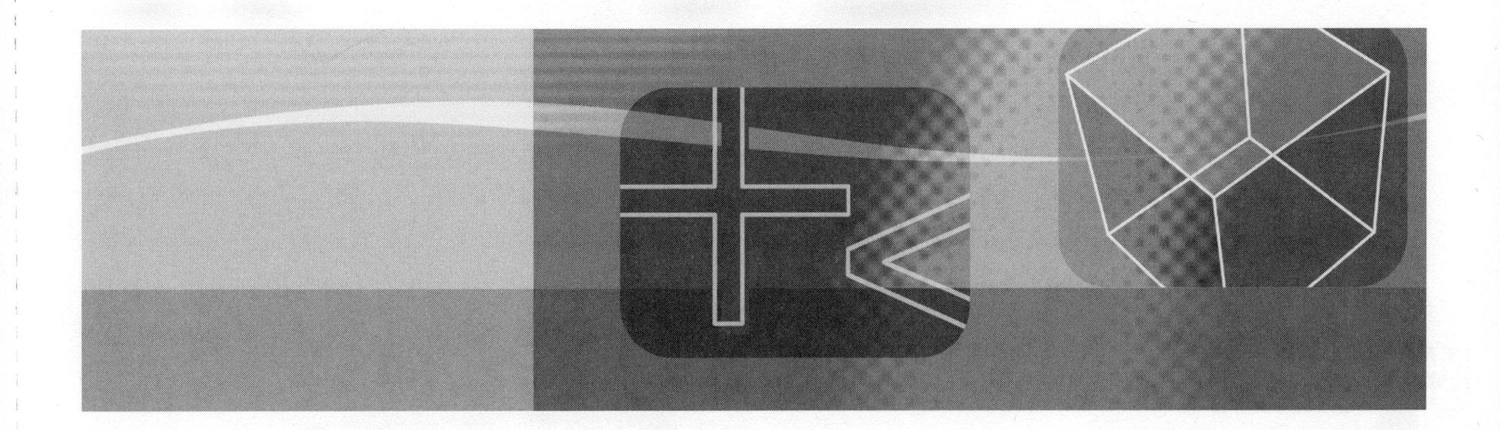

New York, New York   Columbus, Ohio   Chicago, Illinois   Peoria, Illinois   Woodland Hills, California

The McGraw-Hill Companies

Send all inquiries to:
Glencoe/McGraw-Hill
8787 Orion Place
Columbus, OH 43240-4027

ISBN: 0-07-829951-9

Middle School Intervention
Student Workbook

6 7 8 9 10 009 09 08 07 06 05 04

Table of Contents

Name ______________________ Date __________

Large Numbers

A light-year is the distance that light travels in one year. It is a measurement used in astronomy. It is approximately equal to 9,460,000,000,000 kilometers.

EXAMPLE ***Read the number of kilometers in a light-year.***

Trillions			Billions			Millions			Thousands			Ones		
Hundred Trillions	Ten Trillions	Trillions	Hundred Billions	Ten Billions	Billions	Hundred Millions	Ten Millions	Millions	Hundred Thousands	Ten Thousands	Thousands	Hundreds	Tens	Ones
		9,	4	6	0,	0	0	0,	0	0	0,	0	0	0

There are nine trillion, four hundred sixty billion kilometers in one light year.

EXERCISES ***Write each number in words.***

1. 41,020

2. 3,066

3. 4,800,050

4. 78,500,080,000

5. 6,555,800,090,001

6. 80,450,007,000,000

APPLICATIONS ***The chart below shows the distances from the sun to various planets. Use this information to answer Exercises 7–12.***

	Distance from the Sun in Kilometers		
	Farthest	**Nearest**	**Mean**
Mercury	69,000,000	46,000,000	57,900,000
Venus	109,000,000	107,500,000	108,230,000
Earth	152,106,000	147,103,000	149,597,870
Mars	249,100,000	206,500,000	227,900,000
Jupiter	816,000,000	716,000,000	778,000,000
Saturn	1,503,000,000,000	1,351,000,000,000	1,427,000,000,000

7. Write Mercury's nearest distance from the sun in words.

8. Write Jupiter's mean distance from the sun in words.

9. Write Earth's farthest distance from the sun in words.

10. Write Saturn's nearest distance from the sun in words.

11. Write Saturn's farthest distance from the sun in words.

12. Write Jupiter's nearest distance from the sun in words.

SKILL 2

Name ____________________ Date __________

Adding Whole Numbers

In 1950, the number of women in the labor force was 18,408,000. By 1990, the number of women had risen by 38,146,000.

EXAMPLE ***How many women were in the labor force by 1990?***

To find the number of women in the labor force in 1990, add 18,408,000 and 38,146,000.

$$\begin{array}{r} {}^{1}\ \ {}^{1}\qquad\qquad \\ 18{,}408{,}000 \\ +\ 38{,}146{,}000 \\ \hline 56{,}554{,}000 \end{array}$$

There were 56,554,000 women in the labor force in 1990.

EXERCISES *Add.*

1. $\begin{array}{r} 5{,}823 \\ +\ 3{,}246 \\ \hline \end{array}$

2. $\begin{array}{r} 3{,}298 \\ +\ 2{,}049 \\ \hline \end{array}$

3. $\begin{array}{r} 6{,}934 \\ +\ 6{,}841 \\ \hline \end{array}$

4. $\begin{array}{r} 2{,}910 \\ +\ 1{,}208 \\ \hline \end{array}$

5. $\begin{array}{r} 8{,}592 \\ +\ 6{,}228 \\ \hline \end{array}$

6. $\begin{array}{r} 5{,}249 \\ +\ 1{,}004 \\ \hline \end{array}$

7. $\begin{array}{r} 5{,}023 \\ 2{,}285 \\ +\ 1{,}529 \\ \hline \end{array}$

8. $\begin{array}{r} 4{,}298 \\ 5{,}841 \\ +\ 6{,}294 \\ \hline \end{array}$

9. $\begin{array}{r} 8{,}914 \\ 846 \\ +\ 2{,}912 \\ \hline \end{array}$

10. $\begin{array}{r} 1{,}429 \\ 3{,}342 \\ +\ 5{,}130 \\ \hline \end{array}$

11. $\begin{array}{r} 5{,}219 \\ 1{,}712 \\ +\ 4{,}458 \\ \hline \end{array}$

12. $\begin{array}{r} 7{,}131 \\ 922 \\ +\ 1{,}165 \\ \hline \end{array}$

13. 12,890 + 3,348 **14.** 37,890 + 14,529 **15.** 72,917 + 2,847

16. 14,888 + 7,491 **17.** 9,399 + 5,094 **18.** 7,930 + 892

APPLICATIONS ***The chart below gives enrollment information for Springfield Middle School. Use this information to answer Exercises 19–25.***

Springfield Middle School		
Class	**Girls**	**Boys**
6th	165	182
7th	194	177
8th	185	179

19. How many students are in the sixth grade class?

20. How many students are in the seventh grade class?

21. How many students are in the eighth grade class?

22. How many girls are enrolled at Springfield Middle School?

23. How many boys are enrolled at Springfield Middle School?

24. Which grade level has the most students?

25. Are there more girls or more boys attending Springfield Middle School?

26. Patrick has $350 to purchase new ski equipment. He wants to buy a pair of skis for $189, a pair of boots for $78, and a set of ski poles for $35. Does Patrick have enough money to buy the items he wants?

SKILL 3

Name ______________________ Date ____________

Subtracting Whole Numbers

The maximum depth of the Pacific Ocean is 11,034 meters, and the maximim depth of the Atlantic Ocean is 9,219 meters.

EXAMPLE ***How much deeper is the maximum depth of the Pacific Ocean than the maximum depth of the Atlantic Ocean?***

To find the difference between the maximum depths, subtract 9,219 from 11,034.

```
  10  2
 1̸1,10 3̸14
− 9, 2 1 9
  1, 8 1 5
```

The deepest part of the Pacific Ocean is 1,815 meters deeper than the deepest part of the Atlantic Ocean.

EXERCISES ***Subtract.***

1. 678 − 234

2. 7,842 − 5,854

3. 12,600 − 7,343

4. 44,009 − 23,563

5. 15,831 − 6,316

6. 71,808 − 23,545

7. 32,987 − 14,963

8. 201,896 − 192,345

9. 800,000 − 65,125

10. 506,782 − 345,236

11. 172,650 − 68,347

12. 893,765 − 149,477

13. 72,564 − 13,867

14. 66,125 − 54,898

15. 14,671 − 9,655

16. 42,900 − 13,897

17. 216,191 − 3,766

18. 56,002 − 13,999

19. 33,333 − 18,965

20. 70,980 − 35,762

21. 34,782 − 987

APPLICATIONS ***The areas of the Great Lakes are listed at the right. Use this information to answer Exercises 22–25.***

Lake	Area in Square Miles
Superior	31,700
Huron	23,000
Michigan	22,300
Erie	9,910
Ontario	7,540

22. How many more square miles are covered by Lake Superior than Lake Erie?

23. How many more square miles are covered by Lake Huron than by Lake Ontario?

24. How many more square miles are covered by Lake Michigan than by Lake Erie?

25. What is the difference in the area of the Great Lake with the greatest area and the Great Lake with the least area?

26. Lake Erie is the shallowest of the Great Lakes. Its maximum depth is 210 feet. The largest and deepest of the Great Lakes is Lake Superior. Its maximum depth is 1,333 feet. What is the difference between the maximum depths of these two lakes?

Name ______________________ Date __________

Adding and Subtracting Whole Numbers

The island of Hispaniola is divided into two countries, the Dominican Republic and Haiti. The Dominican Republic covers 48,380 square kilometers and Haiti covers 27,750 square kilometers.

EXAMPLES ***What is the total area of the island of Hispaniola?***

$$\begin{array}{r} {\scriptstyle 11\ \ 1\ \ \ } \\ 48{,}380 \\ +\ 27{,}750 \\ \hline 76{,}130 \end{array}$$

The total area of the island is 76,130 square kilometers.

How much more area does the Dominican Republic cover than Haiti?

$$\begin{array}{r} {\scriptstyle 7\ \ \ \ \ \ \ \ } \\ 4\not{8},{}^{1}380 \\ -\ 27{,}750 \\ \hline 20{,}630 \end{array}$$

The Dominican Republic covers 20,630 more square kilometers than Haiti.

EXERCISES ***Add or subtract.***

1. $\begin{array}{r} 785 \\ +\ 361 \\ \hline \end{array}$

2. $\begin{array}{r} 645 \\ -\ 389 \\ \hline \end{array}$

3. $\begin{array}{r} 379 \\ +\ 562 \\ \hline \end{array}$

4. $\begin{array}{r} 6{,}725 \\ -\ 3{,}089 \\ \hline \end{array}$

5. $\begin{array}{r} 7{,}740 \\ +\ 3{,}659 \\ \hline \end{array}$

6. $\begin{array}{r} 12{,}450 \\ -\ 5{,}237 \\ \hline \end{array}$

7. 56,342
+ 8,945

8. 24,893
− 7,934

9. 5,678
+ 37,655

10. 70,843
− 36,900

11. 953,560
− 257,425

12. 73,000
− 13,896

13. 356
275
+ 159

14. 14,893
7,568
+ 56,934

15. 3,855
2,543
+ 980

APPLICATIONS ***The sizes of the world's largest islands are given at the right. Use this information to answer Exercises 16–19.***

Island	Area in Square Miles
Greenland	840,000
New Guinea	306,000
Borneo	280,100
Madagascar	226,658
Baffin	195,928
Sumatra	165,000

16. How much more area does Baffin cover than Sumatra?

17. How much more area does Borneo cover than Madagascar?

18. New Guinea and Borneo are in the South Pacific Ocean. What is the total area of these two islands?

19. What is the total area of the three largest islands?

20. The Nile is the longest river in the world. It is 6,673 kilometers long. The next longest river is the Amazon which is 6,440 kilometers long. How much longer is the Nile than the Amazon?

SKILL 5

Name ______________________ Date __________

Multiplying Whole Numbers

The auditorium at Springdale High School seats 358 people. If all three nights of the school play were sold out, how many tickets were sold?

To find how many tickets, multiply 358 by 3.

STEP 1	STEP 2	STEP 3
Multiply by the ones. $\begin{array}{r} {}^{2} \\ 358 \\ \times\ 3 \\ \hline 4 \end{array}$ *$3 \times 8 = 24$* *Rename 24 ones as 2 tens 4 ones.*	*Multiply by the tens.* $\begin{array}{r} {}^{12} \\ 358 \\ \times\ 3 \\ \hline 74 \end{array}$ *$3 \times 5 = 15$* *$15 + 2 = 17$* *Record 17 tens as shown.*	*Multiply by the hundreds.* $\begin{array}{r} 358 \\ \times\ 3 \\ \hline 1{,}074 \end{array}$ *$3 \times 3 = 9$* *$9 + 1 = 10$*

There were 1,074 tickets sold.

EXERCISES ***Multiply.***

1. $\begin{array}{r} 72 \\ \times\ 3 \\ \hline \end{array}$ **2.** $\begin{array}{r} 92 \\ \times\ 4 \\ \hline \end{array}$ **3.** $\begin{array}{r} \$30 \\ \times\ 6 \\ \hline \end{array}$ **4.** $\begin{array}{r} 242 \\ \times\ 2 \\ \hline \end{array}$

5. $\begin{array}{r} 85 \\ \times\ 4 \\ \hline \end{array}$ **6.** $\begin{array}{r} 79 \\ \times\ 6 \\ \hline \end{array}$ **7.** $\begin{array}{r} 42 \\ \times\ 8 \\ \hline \end{array}$ **8.** $\begin{array}{r} 53 \\ \times\ 7 \\ \hline \end{array}$

9. $\begin{array}{r} 30 \\ \times 16 \\ \hline \end{array}$ **10.** $\begin{array}{r} \$72 \\ \times\ 31 \\ \hline \end{array}$ **11.** $\begin{array}{r} 21 \\ \times\ 10 \\ \hline \end{array}$ **12.** $\begin{array}{r} 64 \\ \times\ 37 \\ \hline \end{array}$

13. $\begin{array}{r} 612 \\ \times 9 \\ \hline \end{array}$

14. $\begin{array}{r} 479 \\ \times 7 \\ \hline \end{array}$

15. $\begin{array}{r} 183 \\ \times 59 \\ \hline \end{array}$

16. $\begin{array}{r} 321 \\ \times 43 \\ \hline \end{array}$

17. $\begin{array}{r} 80 \\ \times 76 \\ \hline \end{array}$

18. $\begin{array}{r} 45 \\ \times 81 \\ \hline \end{array}$

19. $\begin{array}{r} 924 \\ \times 6 \\ \hline \end{array}$

20. $\begin{array}{r} 53 \\ \times 33 \\ \hline \end{array}$

21. 86×5

22. 630×93

APPLICATIONS

23. A satellite in orbit travels 525 kilometers each minute. How many kilometers does it travel in one hour?

24. Mrs. Burns earns $68 each day. How much does she earn during one week in which she works 5 days?

25. A pet shop uses 43 pounds of dog food each day. How many pounds are used in one year?

Name ______________________ Date __________

Dividing Whole Numbers

The sports club expects 295 people to attend the annual awards banquet. Each table seats 12 people. How many tables are needed?

To find how many tables, divide 295 by 12.

STEP 1	STEP 2	STEP 3
Divide the hundreds.	*Divide the tens.*	*Divide the ones.*
12)295 *Since $2 < 12$, the quotient has no hundreds.*	2 12)295 − 24 5	24 R7 12)295 − 24↓ 55 −48 7

The sports club needs 24 tables plus 1 more to seat the remaining 7 people. So, they need 25 tables.

EXERCISES *Divide.*

1. 4)48 **2.** 6)82 **3.** 3)784

4. 3)114 **5.** 8)792 **6.** 8)851

7. 10)70 **8.** 30)900 **9.** 50)370

10. $21\overline{)84}$ 11. $42\overline{)210}$ 12. $61\overline{)854}$

13. $39\overline{)1,530}$ 14. $22\overline{)1,980}$ 15. $8\overline{)9,207}$

16. $539 \div 8$ 17. $14,260 \div 23$

APPLICATIONS

18. Sally bought one gallon of fruit punch for her party. How many 8-ounce servings of punch are in one gallon? (Hint: 1 gallon = 128 ounces)

19. How many dozen dinner rolls are needed to serve 150 people one roll each?

20. About how long does it take to travel 200 miles if you average 65 miles per hour?

Name ______________________ Date __________

Multiplying and Dividing Whole Numbers

A tennis court is 78 feet long.

EXAMPLE ***How long is a tennis court in inches?***

To find the length in inches, multiply 78 by 12.

$$\begin{array}{r} 78 \\ \times\ 12 \\ \hline 156 \\ 78 \\ \hline 936 \end{array}$$

A tennis court is 936 inches long.

EXAMPLE ***How long is a tennis court in yards?***

To find the length in yards, divide 78 by 3.

$$\begin{array}{r} 26 \\ 3\overline{)78} \\ \underline{6} \\ 18 \\ \underline{18} \\ 0 \end{array}$$

A tennis court is 26 yards long.

EXERCISES ***Multiply.***

1. $\begin{array}{r} 28 \\ \times\ 16 \\ \hline \end{array}$

2. $\begin{array}{r} 181 \\ \times\ 15 \\ \hline \end{array}$

3. $\begin{array}{r} 301 \\ \times\ 26 \\ \hline \end{array}$

4. $\begin{array}{r} 261 \\ \times 77 \\ \hline \end{array}$

5. $\begin{array}{r} 450 \\ \times 135 \\ \hline \end{array}$

6. $\begin{array}{r} 713 \\ \times 230 \\ \hline \end{array}$

7. $\begin{array}{r} 709 \\ \times 235 \\ \hline \end{array}$

8. $\begin{array}{r} 1,467 \\ \times 592 \\ \hline \end{array}$

9. $\begin{array}{r} 5,316 \\ \times 1,633 \\ \hline \end{array}$

Divide.

10. $7\overline{)112}$

11. $8\overline{)592}$

12. $5\overline{)2,010}$

13. $12\overline{)528}$

14. $28\overline{)392}$

15. $63\overline{)13,986}$

16. $82\overline{)18,942}$

17. $57\overline{)14,364}$

18. $231\overline{)94,710}$

APPLICATIONS

Use the price list for pizzas to answer Exercises 19–21.

Paul's Pizza	
Small	$11
Medium	$13
Large	$15

19. What is the cost of 3 medium pizzas?

20. The refreshment committee for the class party ordered 27 large pizzas. What is the cost of the pizzas?

21. Which would cost more, 27 large pizzas or 32 medium pizzas?

22. The trip from Albuquerque, New Mexico to Charlotte, North Carolina is 1,624 miles. Hauling a 2,400-pound load, truck driver Jim Cronin wants to make the trip in 4 days. How many miles a day should he travel?

SKILL 8

Name ______________________ Date __________

Order of Operations

Follow the **order of operations** to evaluate or find the value of an expression.

Order of Operations
1. Do all operations within grouping symbols. Start with the innermost grouping symbol. 2. Do all multiplication and division in order from left to right. 3. Do all addition and subtraction in order from left to right.

EXAMPLE ***Evaluate $15 - 2 \times 4 - (6 - 3)$.***

$15 - 2 \times 4 - (6 - 3) = 15 - 2 \times 4 - 3$ *Do all the operations within grouping symbols.*

$= 15 - 8 - 3$ *Do multiplication and division from left to right.*

$= 4$ *Do addition and subtraction from left to right.*

EXERCISES ***Evaluate each expression.***

1. $(5 + 3) \div 2 + 2$

2. $7 \times 5 - 3 \times 4$

3. $3 \times 6 + 9 \div 3 - 6$

4. $5 \times 13 - 8 \times 5 + 6$

5. $(3 + 6) \div 3 \times 3$

6. $(8 - 3)(12 \div 4) - 5$

7. $[36 - (2 + 4)2]3$

8. $20 \div 4 \times 5 \times 2 \div 10$

9. $125 \div [5(2 + 3)]$

10. $2 \times 8 - 42 \div 7 + 4$

11. $150 \div 10 - 3 \times 5$

12. $4 + (5 - 3)5 + 7$

13. $(16 + 4) \div 4$

14. $56 \div [(5 - 3) \times 4]$

15. $15 - 2 \times 4 - (6 - 3)$

16. $[1 + (5 - 2)] \times 6$

17. $2 + 4 \times 9 \div 12$

18. $45 - 40 + 1 \times 2$

19. $(100 - 25) \times 2 + 25$

20. $(12 - 8) \div 4 + 6$

APPLICATIONS ***Use the prices at the right to write mathematical expressions for each total cost. Evaluate the expression to find the total cost.***

Riverview Theater Prices	
Adult	$6.00
Student	$3.00
Senior Citizen	$5.00

21. 3 adult tickets and 4 student tickets

22. 2 adult tickets, 1 senior-citizen ticket, and 3 student tickets

23. 5 adult tickets and 3 senior-citizen tickets

24. 1 senior-citizen ticket and 5 student tickets with a coupon for $2 off the total purchase

25. 5 adult tickets and 2 student tickets with a coupon for $1 off each adult ticket

26. 4 adult tickets and 6 student tickets on a night that offers the special deal that a free ticket is given with each ticket purchased

SKILL 9

Name ______________________ Date __________

Adding Decimals

To add decimals, line up the decimal points. Then add the same way you add whole numbers.

EXAMPLE ***Add 6.22 + 7.4 + 0.895 + 13.***

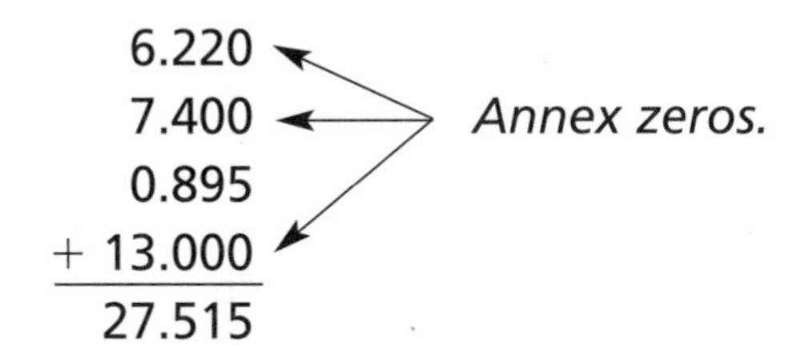

The sum of the numbers is 27.515.

EXERCISES ***Add.***

1. 8.67 + 1.58

2. 13.5 + 26.7

3. 8.476 + 5.72

4. 709.8 + 296.75

5. 6.793 + 15.6

6. 0.058 + 0.48

7. 6.89 + 7.2 + 8.67

8. 12.8 + 8.45 + 34.9

9. 0.78 + 3.7 + 1.666

10. 102.8 + 98.35 + 115.4

11. 0.7 + 11.2 + 8.75

12. 77.85 + 16.1 + 22.48

13. 4.56 + 22.7

14. 3.75 + 8.9

15. 0.97 + 1.9

16. 155.3 + 46.79

17. 35.98 + 4.7 + 37.23

18. 5.68 + 0.9887 + 1.354

19. 56.8 + 4.36 + 1.98 + 2.6

20. 78.91 + 3.476 + 5.65 + 24.8

APPLICATIONS

21. A cyclist with a mass of 58.2 kilograms steps onto a balance scale wearing clothing and a helmet that have a mass of 1.32 kilograms. What is the total mass?

22. A gift box of fruit has 1.4 kilograms of pears, 1.235 kilograms of apples, and 1 kilogram of oranges. What is the total mass of the fruit?

23. Tomo wants to put a decorative border around a triangular flower garden. The lengths of the sides of the garden are 4.36 meters, 3.5 meters, and 5.75 meters. What is the perimeter of the flower garden?

24. A ticket to a movie theater costs $5.25. A large lemonade costs $3, and a small popcorn costs $1.50. What is the total cost of going to the movie and buying a large lemonade and a small popcorn?

25. Rita has a part-time job. On Monday, she worked 3.5 hours. On Tuesday, she worked 4 hours. She did not work Wednesday or Thursday, but she worked 2 hours on Friday and 6.5 hours on Saturday. If she did not work on Sunday, how many hours did she work that week?

26. A barometer rose 3.2 inches in one hour and another 2 inches the next hour. What was the total rise in the barometer?

SKILL 10

Name ________________________ Date __________

Subtracting Decimals

To subtract decimals, line up the decimal points. Then subtract the same way you subtract whole numbers.

EXAMPLE ***Subtract 32.5 − 3.465.***

```
  32.500    Annex zeros.
 − 3.465
  29.035
```

The difference of the numbers is 29.035.

EXERCISES ***Subtract.***

1. 3.9 − 1.5

2. 72.1 − 56.7

3. 3.921 − 2.345

4. 6.789 − 3.56

5. 75.2 − 14.85

6. 6.921 − 1.156

7. 10.34 − 4.8

8. 0.897 − 0.6685

9. 9.03 − 2.8

10. 40 − 13.65

11. 5.72 − 3.9

12. 82.965 − 6.39

13. 25.1 − 3.657

14. 9.871 − 3.9

15. 7.19 − 0.653

16. 6.9 − 2.654

17. 18.564 − 5.8

18. 0.978 − 0.5

19. 4 − 0.875

20. 75.92 − 8.921

21. 72.5 − 61.65

22. 1.872 − 0.98

23. 508 − 8.32

APPLICATIONS

24. One week, the price of gasoline was $1.245 per gallon. The next week, the price of gasoline was $1.269. What was the increase for each gallon of gasoline?

25. One year, the cost of one kilowatt-hour of electricity was $0.094. Two years later, the cost of one kilowatt-hour of electricity was $0.112. What was the increase of each kilowatt-hour of electricity?

26. Herman has a part-time job. Last week, he worked 16 hours. This week, he worked 13.5 hours. How many more hours did he work last week than this week?

27. In 1984, Carl Lewis of the United States won the Olympic gold medal for the 100-meter dash. He ran it in 9.99 seconds. In the 1988 Olympics, he ran the 100-meter dash in 0.07 second less than his Olympic time in 1984. What was his time for the 100-meter dash in the 1988 Olympics?

28. Eduardo has $45. If he buys a CD for $13.98, how much will Eduardo have left?

29. Rita is a gymnast. In one meet, she received 32.45 points. The next meet, she received 1.6 less points. How many points did she receive in the second meet?

Name ______________________ Date __________

Estimating Sums and Differences

Estimation is a useful skill that provides a quick and easy answer when an exact answer is not necessary. Estimation also allows you to check the reasonableness of an answer.

EXAMPLE ***The Spanish Club needs to earn $250 for their annual fiesta. They made $95.50 at a bake sale, $54.90 at a car wash, and $82.35 at an arts and crafts fair. Have they earned enough money? If not, about how much more do they need to earn?***

Estimate the total amount they have earned by rounding each amount to the nearest ten. Then add.

$$\begin{array}{rcr} \$95.50 & \rightarrow & \$100 \\ 54.90 & \rightarrow & 50 \\ +\ 82.35 & \rightarrow & +\ 80 \\ \hline & & \$230 \end{array}$$

The club has earned about $230 which is not enough for the fiesta. Since $250 − $230 = $20, they still need to earn about $20.

EXERCISES ***Estimate each sum or difference.***

1. 336 + 652 **2.** 8,924 − 814 **3.** 2,347 + 906

4. 12,453 + 510 **5.** 23,200.7 + 14,297.5 **6.** 56,231 − 4,801

7. 34,412.4 − 16,250.5 **8.** 98 + 42 + 76 **9.** 176 + 984 + 16

10. \$13.72 − \$5.19 **11.** \$456.50 − \$299.17 **12.** \$25.50 + \$42.99

13. \$114 + \$56 + 74 **14.** \$3,679 + \$34.50 **15.** \$15,099.99 − \$3,500

APPLICATIONS ***The data in the table below shows the makeup of the population of several cities in the United States according to the 1990 census. Use the data to answer Exercises 16–20.***

City	Females	Males
Boston, Massachusetts	298,311	275,972
Chicago, Illinois	1,449,021	1,334,705
El Paso, Texas	268,179	247,163
Omaha, Nebraska	175,403	160,392
Phoenix, Arizona	495,814	487,589

16. Estimate the total population of Chicago, Illinois.

17. Which city has about twice the total population of El Paso, Texas?

18. About how many more females than males live in Boston, Massachusetts?

19. Which city has about 15,000 more females than males?

20. Suppose the female population of El Paso, Texas decreases by about 10,000 and the male population increases by about 10,000. How would the female and male population now compare?

21. Andrea started with a balance of \$450.98 in her checking account. She then wrote checks for \$124.12, \$4.68, and \$39.50. She then made a deposit of \$213.95. Estimate the balance in her account after these transactions.

Name ______________________ Date __________

Multiplying Whole Numbers and Decimals

EXAMPLES

Multiply 182 by 51.

$$\begin{array}{r} 182 \\ \times\ 51 \\ \hline 182 \\ 910 \\ \hline 9{,}282 \end{array}$$

The product is 9,282.

Multiply 8.4 by 0.62.

$$\begin{array}{rl} 8.4 & \leftarrow \textit{1 decimal place} \\ \times\ 0.62 & \leftarrow \textit{2 decimal places} \\ \hline 168 & \\ 5\ 04 & \\ \hline 5.208 & \leftarrow \textit{3 decimal places} \end{array}$$

The sum of the decimal places in the factors is 3, so the product has 3 decimal places.

The product is 5.208.

EXERCISES

Multiply.

1. $\begin{array}{r} 147 \\ \times\ 6 \\ \hline \end{array}$

2. $\begin{array}{r} 63 \\ \times\ 51 \\ \hline \end{array}$

3. $\begin{array}{r} 182 \\ \times\ 51 \\ \hline \end{array}$

4. $\begin{array}{r} 62 \\ \times\ 12 \\ \hline \end{array}$

5. $\begin{array}{r} 5.84 \\ \times\ 0.08 \\ \hline \end{array}$

6. $\begin{array}{r} 0.33 \\ \times\ 6.5 \\ \hline \end{array}$

7. $\begin{array}{r} 2.48 \\ \times\ 0.66 \\ \hline \end{array}$

8. $\begin{array}{r} 0.55 \\ \times\ 1.7 \\ \hline \end{array}$

9. $\begin{array}{r} 1.2 \\ \times\ 0.003 \\ \hline \end{array}$

10. $\begin{array}{r} 0.52 \\ \times\ 0.03 \\ \hline \end{array}$

11. $\begin{array}{r} 29.1 \\ \times\ 0.29 \\ \hline \end{array}$

12. $\begin{array}{r} 0.0054 \\ \times\ 6.1 \\ \hline \end{array}$

APPLICATIONS *Tonya is reading a map. Use the scale below to answer Exercises 13–15.*

Map Scale
1 centimeter = 34 kilometers

13. What is the distance represented by 16 centimeters on the map?

14. What is the distance represented by 7.4 centimeters on the map?

15. What is the distance represented by 12.8 centimeters on the map?

16. On the average, 130 words are listed on one page of a dictionary. How many words would you expect to be listed on 520 pages?

17. During his professional basketball career, Wilt Chamberlain averaged about 30.06 points per game for 1,045 games. How many points did he score in his career?

18. Stewart buys 2.8 pounds of steak. If the steak costs $5.70 per pound, what is the total cost of the steak?

19. The speed of the spinetailed swift has been measured at 106.25 miles per hour. At that rate, how far can it travel in an hour and a half?

SKILL 13

Name ______________________ Date __________

Dividing Decimals

EXAMPLES ***Divide 54.4 by 17.***

$$17\overline{)54.4} = 3.2$$

```
     3.2
17)54.4
   51
    3 4
    3 4
      0
```

Divide as with whole numbers, placing the decimal point above the decimal point in the dividend.

The quotient is 3.2.

Divide 0.5194 by 0.49.

```
          1.06
0.49.)0.51.94
        49
         2 9
           0
         2 94
         2 94
            0
```

Change 0.49 to 49 by moving the decimal point two places to the right.

Move the decimal point in the dividend the same number of places to the right.

Divide as with whole numbers placing the decimal point above the new point in the dividend.

The quotient is 1.06.

EXERCISES ***Divide.***

1. $5\overline{)125}$

2. $8\overline{)992}$

3. $24\overline{)43.2}$

4. $11\overline{)3.091}$

5. $3\overline{)3.066}$

6. $2.4\overline{)0.192}$

7. $0.3\overline{)129}$

8. $0.44\overline{)52.8}$

9. $4.5\overline{)40.05}$

10. $0.3\overline{)3.066}$ 11. $4.5\overline{)40.05}$ 12. $11\overline{)30.91}$

13. $1.4\overline{)121.8}$ 14. $8\overline{)0.0092}$ 15. $0.38\overline{)760.38}$

APPLICATIONS ***Herman's Farm Market lists its prices below. Use this information to answer Exercises 16–18.***

Herman's Farm Market	
Tomatoes	3 pounds for $2.16
Corn	1 dozen for $4.20
Potatoes	5 pounds for $2.15

16. What is the price of tomatoes per pound?

17. What is the price of the potatoes per pound?

18. What is the price for one ear of corn?

19. Sue earns $195.20 in a week in which she works 30.5 hours. What is Sue's hourly pay rate?

20. Three friends plan to divide the cost of a birthday gift for another friend. If the cost of the gift is $16.38, what is each person's share of the cost?

21. What is the cost of a gallon of gasoline at a gasoline station where a sale of 6.8 gallons costs a customer $9.18?

22. A dolphin can swim at a speed of about 37 miles per hour. The fastest human swimmer can reach a speed of about 5.2 miles per hour. About how many times faster than humans are dolphins?

Name ______________________ Date __________

Estimating Products and Quotients

You can use estimation to find a quick and easy answer when an exact answer is not necessary.

EXAMPLE ***Myron bought 19.867 gallons of gas at $1.19 a gallon. About how much did he pay for the gas?***

One way to estimate the cost is to round 19.867 to the nearest ten and $1.19 to the nearest dime. Then multiply.

$$\begin{array}{rcr} 19.867 & \rightarrow & 20 \\ \underline{\times\ 1.19} & \rightarrow & \underline{\times\ 1.20} \\ & & 24.00 \end{array}$$

The gas cost about $24.00.

EXERCISES ***Estimate each product or quotient.***

1. 45×6

2. 89×32

3. $58 \div 7$

4. $752 \div 8$

5. 128×54

6. 36.9×8

7. $792 \div 92$

8. $41.2 \div 8$

9. 4.8×19.7

10. $\$18.36 \times 12$

11. $33 \div 7.7$

12. 27.5×24.6

13. $\$15.35 \div 4$

14. 59.1×2.09

15. $27.26 \div 2.6$

16. 55.8 ÷ 2.29 **17.** 72.9 × 3.09 **18.** 88.1 ÷ 29.2

19. Tina says that when she uses a calculator to solve problems, she does not need to estimate her answer first. Do you agree? Why or why not?

APPLICATIONS ***In 1993, the tax rate on gross incomes between $36,900 and $89,150 on tax returns filed jointly was 28%. Use this information to answer Exercises 20–23. (HINT: 28% = 0.28)***

20. Describe what you would do to estimate the taxes on gross incomes between $36,900 and $89,150.

21. Estimate the taxes on the lowest gross income in this tax bracket.

22. Estimate the taxes on the highest gross income in this tax bracket.

23. Why might you want to estimate your taxes before determining your exact tax?

24. When Leon visited Mexico, $1 in American money could be exchanged for 2,702.7 pesos. About how many pesos did Leon receive for $28 in American money?

Name ______________________ Date __________

Multiplying by Powers of Ten

The exponent in a power of ten is the same as the number of zeros in the number.

Powers of Ten	
10^0	1
10^1	10
10^2	100
10^3	1,000
10^4	10,000
10^5	100,000

To multiply by a power of ten, move the decimal point to the right the number of places shown by the exponent or the number of zeros. Annex zeros if necessary.

EXAMPLES ***Find each product.***

0.08×10^4

$0.0800 = 800$ *Move the decimal point 4 places to the right.*

The product is 800.

$6.25 \times 1{,}000$

$6.250 = 6{,}250$ *Move the decimal point 3 places to the right.*

The product is 6,250.

EXERCISES ***Choose the correct product.***

1. 2.48×100; 0.0248 or 248

2. 0.9×10^0; 9 or 0.9

3. 0.039×10^2; 3.9 or 39

4. 1.5×10^4; 150,000 or 15,000

Multiply.

5. 15.24×10

6. 0.702×100

7. $5.149 \times 1{,}000$

8. 0.52×100

9. 2.587×10^0

10. 0.2674×100

11. 6.8×10^2

12. 9.57×10^4

13. 6.2×10^5

Solve each equation.

14. $d = 0.92 \times 100$

15. $12.43 \times 10^3 = h$

16. $h = 3.68 \times 10^6$

17. $a = 0.004 \times 10^2$

18. $0.23 \times 1{,}000 = j$

19. $1.89 \times 10^0 = v$

20. $d = 10{,}000 \times 7.07$

21. $0.014 \times 10^2 = k$

22. $v = 589 \times 10^1$

APPLICATIONS

23. What is the length of the Amazon river if it can be represented by 3.9×10^3 miles long? How much longer is it than the Wood River which is 5.7×10^2?

24. The United States spends 37.3×10^9 dollars on research and development in the military. Germany spends 1.4×10^9 dollars on research and development in the military. How much money do these two countries spend altogether?

25. The diameter of Neptune is about 4.95×10^4 kilometers. The diameter of Venus is about 1.21×10^4 kilometers. About how much greater is Neptune's diameter?

Name ______________________ Date __________

Dividing by Powers of Ten

The exponent in a power of ten is the same as the number of zeros in the number.

Powers of Ten

10^0	1
10^1	10
10^2	100
10^3	1,000
10^4	10,000
10^5	100,000

To divide by a power of ten, move the decimal point to the left the number of places shown by the exponent or the number of zeros.

EXAMPLES ***Find each quotient.***

$8 \div 10^4 = 0.0008$ *Move the decimal point 4 places to the left.*

The quotient is 0.0008.

$62.5 \div 1{,}000 = 0.0625$ *Move the decimal point 3 places to the left.*

The quotient is 0.0625.

EXERCISES ***Choose the correct quotient.***

1. $2.48 \div 100$; 0.0248 or 248

2. $0.9 \div 10^0$; 9 or 0.9

3. $0.39 \div 10^2$; 0.039 or 0.0039

4. $1.5 \div 10^4$; 0.00015 or 15,000

Divide.

5. $15.24 \div 10$

6. $0.702 \div 100$

7. $514.9 \div 1{,}000$

8. $5.2 \div 100$ **9.** $2.587 \div 10^0$ **10.** $267.4 \div 100$

11. $68 \div 10^2$ **12.** $9.57 \div 10^4$ **13.** $6{,}245 \div 10^5$

Solve each equation.

14. $d = 92 \div 100$ **15.** $12.43 \div 10^3 = h$ **16.** $h = 36.8 \div 10^6$

17. $a = 0.004 \div 10^2$ **18.** $2{,}358 \div 1{,}000 = j$ **19.** $1.89 \div 10^0 = v$

20. $d = 76.9 \div 10{,}000$ **21.** $8{,}714 \div 10^2 = k$ **22.** $v = 589 \div 10^1$

APPLICATIONS

23. Mr. Fraley bought 1,000 postage stamps for $290 for use in his office. How much did each stamp cost?

24. Mary donated 100 cans of soup to the local food pantry. It cost her $23 to buy the soup. How much did each can of soup cost?

25. George has $245.60 that he wants to split evenly with his 10 nieces and nephews. How much money will each one receive?

26. The planet Saturn is an average distance of about 887,000,000 miles from the sun. If a space ship could travel that distance in 10,000 hours, how fast would it be going?

Name ______________________ Date __________

Divisibility Rules

Sometimes we need to know if a number is **divisible** by another number. In other words, does a number divide evenly into another number. You can use divisibility rules.

A number is divisible by:

- 2 if the ones digit is divisible by 2.
- 3 if the sum of the digits is divisible by 3.
- 5 if the ones digit is 0 or 5.
- 6 if the number is divisible by 2 and 3.
- 9 if the sum of the digits is divisible by 9.
- 10 if the ones digit is zero.

EXAMPLE ***Determine whether 2,346 is divisible by 2, 3, 5, 6, 9, or 10.***

2: The ones digit is 6 which is divisible by 2.
So 2,346 is divisible by 2.

3: The sum of the digits $(2 + 3 + 4 + 6 = 15)$ is divisible by 3.
So 2,346 is divisible by 3.

5: The ones digit is *not* 0 or 5.
So 2,346 is *not* divisible by 5.

6: The number is divisible by 2 and 3.
So 2,346 is divisible by 6.

9: The sum of the digits $(2 + 3 + 4 + 6 = 15)$ is *not* divisible by 9.
So 2,346 is *not* divisible by 9.

10: The ones digit is not 0.
So 2,346 is *not* divisible by 10.

2,346 is divisible by 2, 3, and 6.

EXERCISES ***Use the divisibility rules to determine whether the first number is divisible by the second number.***

1. 3,465,870; 5

2. 5,653,121; 3

3. 34,456,433; 9

4. 6,432; 10

5. 42,981; 2

6. 73,125; 3

7. 3,469; 6

8. 3,522; 6

Determine whether each number is divisible by 2, 3, 5, 6, 9, or 10.

9. 660

10. 5,025

11. 5,091

12. 356

13. 240

14. 657

15. 8,760

16. 3,408

17. 4,605

18. 7,800

19. 8,640

20. 432

21. 1,700,380

22. 4,937,728

APPLICATIONS

23. Ms. Vescelius wants to divide her class into cooperative learning groups. If there are 28 students in the class and she wants all the groups to have the same number of students, how many students should she put in each group?

24. The Kennedy High School band has 117 members. The band director is planning rectangular formations for the band. What formations could he make with all the band members?

25. Fisher Mountain Bike Company wants to produce between 1,009 and 1,030 mountain bicycles per month. Since the demand for the bicycles is great everywhere, they want to ship equal numbers to each of their 6 stores. Find the possible number of bicycles Fisher should ship.

26. Name the greatest 4-digit number that is divisible by 2, 3, and 5.

Name ______________________ Date __________

Prime Factorization

A **prime number** is a whole number greater than 1 that has exactly two factors, 1 and itself.

EXAMPLE ***Name two prime numbers.***

7 factors: 1, 7
23 factors: 1, 23

Therefore, 7 and 23 are prime numbers.

A **composite number** is a whole number greater than 1 that has more than two factors. Every composite number can be written as the product of prime numbers. This is called the **prime factorization** of the number.

EXAMPLE ***Write the prime factorization of 420.***

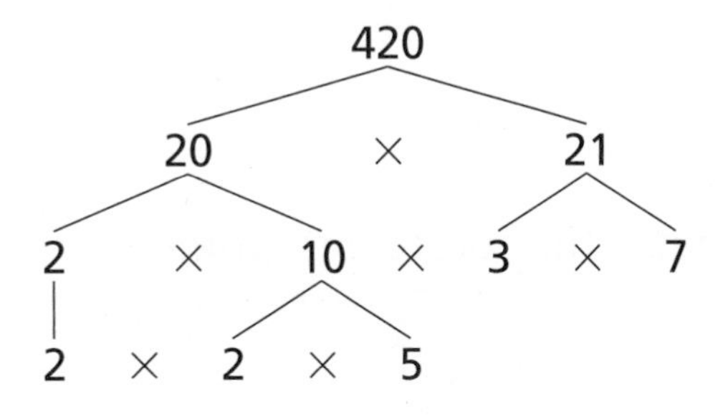

Write 420 as the product of two factors. Keep factoring until all of the factors are prime numbers.

The prime factorization of 420 is $2 \times 2 \times 3 \times 5 \times 7$, or $2^2 \times 3 \times 5 \times 7$.

EXERCISES ***Determine whether each number is composite or prime.***

1. 34 **2.** 77 **3.** 37 **4.** 89

5. 31 **6.** 434 **7.** 97 **8.** 123

Write the prime factorization of each number.

9. 490 **10.** 225 **11.** 900 **12.** 1,105

13. 66 **14.** 306 **15.** 2,475 **16.** 1,024

Find the missing factor.

17. $3^2 \times 5 \times$ ______ $= 315$

18. $2^4 \times$ ______ $\times 7 = 1{,}008$

19. $3^3 \times$ ______ $= 135$

20. $2^2 \times 3^2 \times$ ______ $= 252$

21. $5^2 \times$ ______ $= 275$

22. $3^3 \times 5^2 \times$ ______ $= 7{,}425$

APPLICATIONS

23. The first prime number is 2. What is the fourteenth prime number?

24. Two and 3 are consecutive prime numbers. Why aren't there any other pairs of consecutive prime numbers?

25. Evaluate $n^2 + n + 41$ for $n = 0, 1, 2,$ and 3 to find four prime numbers.

Name ______________________ Date __________

Greatest Common Factor

The **greatest common factor (GCF)** of two or more numbers is the greatest number that is a factor of each number. One way to find the GCF is to list the factors of each number and then choose the greatest of the common factors.

EXAMPLE ***Find the GCF of 72 and 108.***

factors of 72: **1**, **2**, **3**, **4**, **6**, 8, **9**, **12**, **18**, 24, **36**, 72
factors of 108: **1**, **2**, **3**, **4**, **6**, **9**, **12**, **18**, 27, **36**, 54, 108

common factors: 1, 2, 3, 4, 6, 9, 12, 18, 36

The GCF of 72 and 108 is 36.

Another way to find the GCF is to write the prime factorization of each number. Then identify all common prime factors and find their product.

EXAMPLE ***Find the GCF of 210 and 525.***

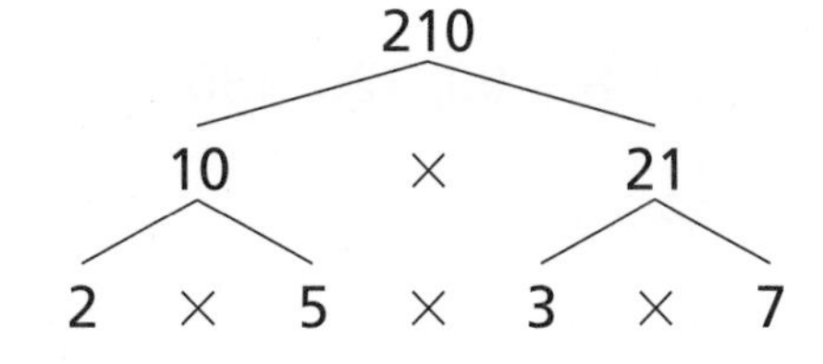

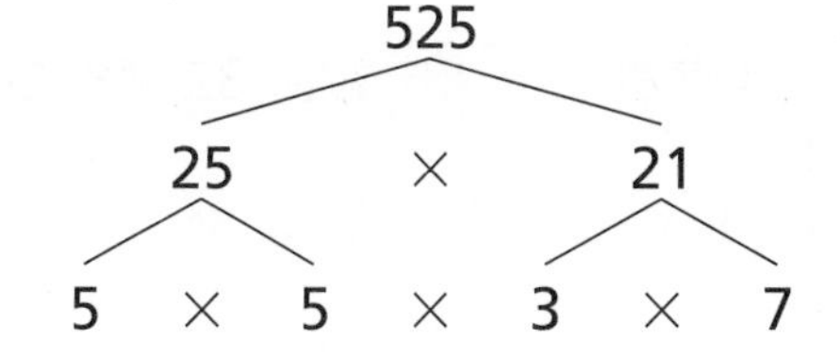

common prime factors: 3, 5, 7

The GCF of 210 and 525 is $3 \times 5 \times 7$, or 105.

EXERCISES ***Find the GCF of each set of numbers by listing the factors of each number.***

1. 12, 18 **2.** 44, 153 **3.** 16, 30

List the common prime factors for each pair of numbers. Then write the GCF.

4. $80 = 2^4 \times 5$
$110 = 2 \times 5 \times 11$

5. $42 = 2 \times 3 \times 7$
$49 = 7 \times 7$

6. $16 = 2^4$
$48 = 2^4 \times 3$

Find the GCF of each pair of numbers by writing the prime factorization of each number.

7. 35, 85

8. 40, 100

9. 42, 23

Find the GCF of each set of numbers.

10. 18, 30

11. 60, 45

12. 24, 72

13. 54, 36

14. 120, 200

15. 81, 153

16. 60, 24, 72

17. 32, 48, 80

18. 90, 120, 180

19. What is the GCF of $2^3 \times 3^2 \times 5$ and $2^2 \times 3^2 \times 5^3$?

APPLICATIONS

20. What is the GCF of all the numbers in the sequence 12, 24, 36, 48, . . .?

21. There are 84 turkey and 63 ham sandwiches to be placed on trays. Each tray should have only one kind of sandwich, and all trays have the same number of sandwiches. What is the greatest number of sandwiches that can be placed on one tray?

Name ______________________ Date __________

Least Common Multiple

A **multiple** of a number is the product of that number and any whole number. The least nonzero multiple of two or more numbers is the **least common multiple (LCM)** of the numbers.

EXAMPLE ***Find the least common multiple of 15 and 20.***

positive multiples of 15: 15, 30, 45, **60**, 75, 90, 105, **120**, . . .
positive multiples of 20: 20, 40, **60**, 80, 100, **120**, 140, . . .

The LCM of 15 and 20 is 60.

Prime factorization can also be used to find the LCM.

EXAMPLE ***Find the LCM of 8, 12, and 18.***

$$8 = 2 \times 2 \times 2$$
$$12 = 2 \times 2 \times 3$$
$$18 = 2 \times 3 \times 3$$
$$2 \times 2 \times 2 \times 3 \times 3 = 72$$

Find prime factors of each number.
Circle all sets of common factors.
Multiply the common factors and any other factors.

The LCM of 8, 12, and 18 is 72.

EXERCISES ***Find the LCM of each set of numbers by listing the multiples of each number.***

1. 12, 16 **2.** 15, 24 **3.** 7, 9

Find the LCM of each set of numbers by writing the prime factorization.

4. 18, 27 **5.** 30, 21 **6.** 20, 50

Find the LCM of each set of numbers.

7. 250, 30 **8.** 8, 54 **9.** 30, 65

10. 6, 10, 15 **11.** 2, 16, 24 **12.** 7, 8, 14

13. 6, 8, 36 **14.** 18, 30, 50 **15.** 14, 22

16. Find the GCF and LCM for 12 and 24.

17. Find the two smallest numbers whose GCF is 9 and whose LCM is 54.

18. List the first four multiples of $2n$.

APPLICATIONS

19. James goes to the zoo every six months, he goes to the art museum every 18 months, and he goes to the children's museum every July 1. This year on July 1, he went to all three places. When will be the next time that he happens to go to all three places?

20. On a store's 100th anniversary, every person who enters gets a pin. Every fourth person gets a mug. Every tenth person gets perfume. Every 25th person gets an umbrella, and every 75th person gets a free dinner. Which shopper will be the first to get all 5 gifts?

SKILL 21

Name ________________________ Date ____________

Simplifying Fractions

There are 30 students in the school chorale, and 12 of these students can stay after school today to help prepare the stage for the concert.

EXAMPLE ***What fraction of the students in chorale can stay after school today? Write the fraction in simplest form.***

From the information, $\frac{12}{30}$ of the students can stay after school.

To simplify this fraction, find the greatest common factor of 12 and 30. The GCF is 6. Then divide the numerator and denominator by 6.

$$\frac{12 \div 6}{30 \div 6} = \frac{2}{5}$$

Therefore, $\frac{2}{5}$ of the students can stay after school.

EXERCISES ***Write each fraction in simplest form.***

1. $\frac{14}{20}$ **2.** $\frac{15}{35}$ **3.** $\frac{16}{20}$

4. $\frac{10}{40}$ **5.** $\frac{16}{36}$ **6.** $\frac{45}{48}$

7. $\frac{22}{55}$ **8.** $\frac{49}{56}$ **9.** $\frac{13}{26}$

10. $\frac{16}{32}$ **11.** $\frac{14}{49}$ **12.** $\frac{60}{80}$

13. $\frac{15}{25}$ **14.** $\frac{16}{18}$ **15.** $\frac{24}{36}$

16. $\frac{8}{32}$ **17.** $\frac{18}{81}$ **18.** $\frac{8}{56}$

19. $\frac{75}{100}$ **20.** $\frac{15}{25}$ **21.** $\frac{4}{44}$

22. $\frac{10}{65}$ **23.** $\frac{28}{63}$ **24.** $\frac{42}{52}$

25. $\frac{25}{150}$ **26.** $\frac{81}{90}$ **27.** $\frac{35}{105}$

APPLICATIONS *Use the data below to answer Exercises 28–35. Write all answers in simplest form.*

1993 U.S. Television Ownership	
Equipment	**Number of Households out of 100**
Television	98
Color Television	97
VCR	80
Two or More Televisions	64
Basic Cable	62
One or More Pay Cable Channels	30
Satellite Dish	4

28. What fraction of U.S. households have a television?

29. What fraction of U.S. households have a color television?

31. What fraction of U.S. households have a VCR?

32. What fraction of U.S. households have two or more televisions?

33. What fraction of U.S. households have basic cable?

34. What fraction of U.S. households have at least one cable channel?

35. What fraction of U.S. households have a satellite dish?

SKILL 22

Name ______________________ Date __________

Mixed Numbers and Improper Fractions

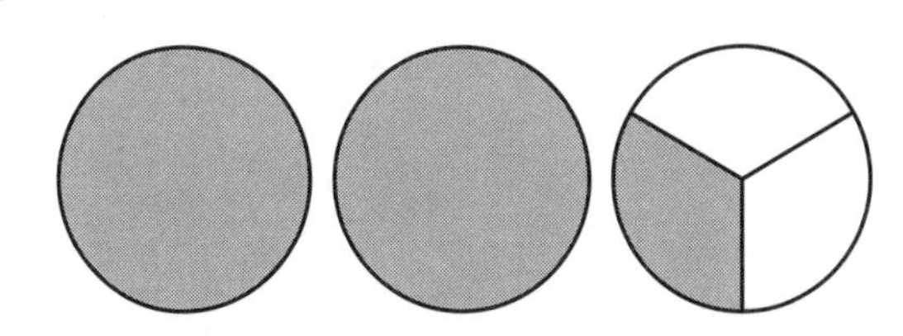

The figure at the right shows 2 whole circles plus $\frac{1}{3}$ of a circle. The **mixed number** $2\frac{1}{3}$ describes the number of circles.

Mixed numbers may be expressed as **improper fractions**. An improper fraction is a fraction in which the numerator is greater than the denominator.

To express a mixed number as an improper fraction, multiply the whole number by the denominator. Add the numerator to the product. Write the sum over the denominator.

EXAMPLE ***Express $2\frac{1}{3}$ as an improper fraction.***

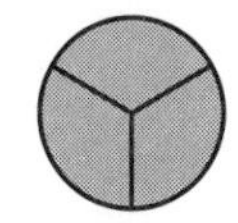
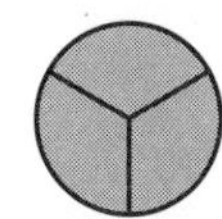
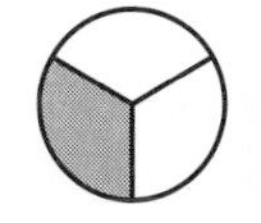

$$2\frac{1}{3} = \frac{(2 \times 3) + 1}{3} = \frac{7}{3}$$

Therefore, $2\frac{1}{3} = \frac{7}{3}$.

An improper fraction may be written as a mixed number.

To express an improper fraction as a mixed number, divide the numerator by the denominator. Write the quotient as the whole number. Write the remainder over the denominator as the fraction.

EXAMPLE ***Express $\frac{7}{4}$ as a mixed number.***

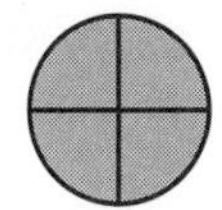
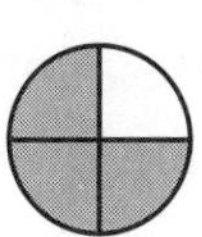

$7 \div 4 = 1 \text{ R } 3$ or $1\frac{3}{4}$

Therefore, $\frac{7}{4} = 1\frac{3}{4}$.

EXERCISES ***Draw a model and express each mixed number as an improper fraction.***

1. $1\frac{3}{8}$ **2.** $2\frac{1}{4}$ **3.** $1\frac{5}{6}$

Draw a model and express each fraction as a mixed number.

4. $\frac{5}{2}$

5. $\frac{9}{5}$

6. $\frac{15}{4}$

Express each mixed number as an improper fraction.

7. $6\frac{1}{2}$

8. $3\frac{7}{8}$

9. $2\frac{8}{9}$

10. $10\frac{2}{3}$

11. $5\frac{4}{7}$

12. $4\frac{5}{6}$

13. $9\frac{1}{4}$

14. $8\frac{3}{5}$

Express each fraction as a mixed number.

15. $\frac{19}{6}$

16. $\frac{27}{4}$

17. $\frac{52}{9}$

18. $\frac{25}{2}$

19. $\frac{37}{5}$

20. $\frac{77}{8}$

21. $\frac{41}{3}$

22. $\frac{31}{7}$

APPLICATIONS

23. Suppose it snowed 5 inches in 2 days. The improper fraction $\frac{5}{2}$ tells the average daily snowfall. Write the improper fraction as a mixed number.

24. The Windsor Bay Deli sold $2\frac{1}{8}$ apple pies on Wednesday.

    If each piece was $\frac{1}{8}$ of a pie, how many pieces of pie were sold?

SKILL 23 Name ____________________ Date __________

Ratios as Fractions

Four adults accompany Mr. Goetz's class on a field trip to the municipal court. There are 27 students going on the field trip.

EXAMPLES ***What is the ratio of adults to students?***

$$\frac{\text{number of adults}}{\text{number of students}} = \frac{4}{27}$$

The ratio is $\frac{4}{27}$.

What is the ratio of students to the total number of people going on the field trip?

There are 4 + 27 or 31 people going on the trip.

$$\frac{\text{number of students}}{\text{total number}} = \frac{27}{31}$$

The ratio is $\frac{27}{31}$.

EXERCISES ***Express each ratio as a fraction.***

1. 30 out of 50 doctors

2. 84 students to 3 teachers

3. 22 players to 2 teams

4. 20 wins in 32 games

5. 4 boys to 6 girls

6. $8 for 2 tickets

7. 14 wins to 35 losses

8. 6 hits to 14 times at bat

9. 8 out of 10 people

10. 90 women to 144 men

11. 8 out of 10 bicycles

12. 5 out of 14 weeks

APPLICATIONS ***Ms. McClure's math class took a survey to determine what types of pets members of the class owned. There are 28 students in the class. Use the data at the right to answer Exercises 13–20.***

Number of Class Members Who Own Pets	
Cats	12
Dogs	11
Fish	9
Birds	5
Other	3
None	2

13. What ratio of the class members own a cat?

14. What ratio of the class members own a fish?

15. What ratio of the class members own a bird?

16. What ratio of the class members do *not* own a dog?

17. What ratio of the class members own a dog or no pet at all?

18. What ratio of the class members do *not* own a bird?

19. What ratio of the class members own a pet?

20. Do some of the class members own more than one pet? Explain.

SKILL 24

Name ______________________ Date __________

Changing Fractions to Decimals

A fraction is another way of writing a division problem. To change a fraction to a decimal, divide the numerator by the denominator.

EXAMPLE ***About $\frac{1}{20}$ of the heat in a house is lost through the doors. Write this fraction as a decimal.***

$\frac{1}{20}$ means $1 \div 20$ or $20\overline{)1}$.

$$\begin{array}{r} 0.05 \\ 20\overline{)1.00} \end{array}$$

So, $\frac{1}{20} = 0.05$.

EXERCISES ***Express each fraction as a decimal. Use bar notation if necessary.***

1. $\frac{4}{25}$

2. $\frac{3}{5}$

3. $\frac{7}{20}$

4. $\frac{3}{50}$

5. $\frac{9}{10}$

6. $\frac{7}{8}$

7. $\frac{1}{3}$

8. $\frac{14}{16}$

9. $\frac{20}{30}$

10. $\frac{5}{9}$

11. $\frac{19}{20}$

12. $\frac{5}{200}$

13. $\frac{10}{50}$ 14. $\frac{13}{20}$ 15. $\frac{5}{6}$ 16. $\frac{4}{5}$

17. $\frac{7}{10}$ 18. $\frac{13}{40}$ 19. $\frac{39}{50}$ 20. $\frac{2}{25}$

21. $\frac{7}{16}$ 22. $\frac{34}{125}$ 23. $\frac{16}{25}$ 24. $\frac{99}{100}$

25. $\frac{17}{20}$ 26. $\frac{3}{150}$ 27. $\frac{3}{8}$ 28. $\frac{2}{3}$

APPLICATIONS ***A mill is a unit of money that is used in assessing taxes. One mill is equal to $\frac{1}{10}$ of a cent or $\frac{1}{1,000}$ of a dollar.***

29. Money is usually written using decimals. Express each fraction above as a decimal using the correct money symbol.

30. Find the number of cents and the number of dollars equal to 375 mills.

31. Find the number of cents and the number of dollars equal to 775 mills.

32. Find the number of cents and the number of dollars equal to 1,000 mills.

SKILL 25

Name ______________________________ Date ____________

Comparing and Ordering Fractions

As of 1992, the New York Yankees had won 22 of the 33 World Series games in which they had played. The St. Louis Cardinals had won 9 of the 15 World Series games in which they had played.

EXAMPLE ***Which team has a better record in the World Series?***

To answer this question, compare $\frac{22}{33}$ and $\frac{9}{15}$.

One way to compare these fractions is to express them as decimals and then compare the decimals.

$\frac{22}{33} = 0.6666666667$ $\frac{9}{15} = 0.6$

Since $0.6666666667 > 0.6$, $\frac{22}{33} > \frac{9}{15}$.

The New York Yankees have the better record.

EXERCISES ***Fill in each □ with $<$, $>$, or $=$ to make a true sentence.***

1. $\frac{2}{7} \square \frac{3}{8}$

2. $\frac{3}{11} \square \frac{1}{5}$

3. $\frac{11}{21} \square \frac{9}{16}$

4. $\frac{14}{21} \square \frac{10}{15}$

5. $\frac{25}{27} \square \frac{17}{19}$

6. $\frac{3}{10} \square \frac{4}{9}$

7. $1\frac{7}{8} \square 1\frac{4}{5}$

8. $3\frac{7}{9} \square 3\frac{6}{7}$

9. $5\frac{10}{19} \square 5\frac{15}{24}$

Write each set of fractions in order from least to greatest.

10. $\frac{3}{5}, \frac{7}{9}, \frac{4}{5}, \frac{1}{2}$

11. $\frac{3}{8}, \frac{2}{7}, \frac{8}{11}, \frac{5}{16}$

12. $\frac{9}{14}, \frac{6}{7}, \frac{3}{4}, \frac{12}{19}$

13. $\frac{11}{23}, \frac{19}{27}, \frac{7}{10}, \frac{15}{17}$

APPLICATIONS ***The Pittsburgh Pirates have won 14 out of 21 games, and the New York Mets have won 15 out of 23 games. Use this information to answer Exercises 14–17.***

14. Which team has the better record?

15. Suppose the Pirates win 2 of their next three games and the Mets win all of their next 3 games. Which team has the better record?

16. Suppose the Pirates went on to win 21 games after playing 30 games. Is their record better now than it was before? Explain.

17. Suppose the Mets went on to win 16 games after playing 30 games. Is their record better now than it was before? Explain.

18. Larry has $\frac{5}{6}$ yard of material. Does he have enough to make a vest that requires $\frac{3}{4}$ yard of material? Explain.

SKILL 26

Name ______________________ Date ____________

Adding Fractions

To add fractions with like denominators, add the numerators. Write the sum over the common denominator. Simplify the sum if possible.

EXAMPLE *Find the sum of $\frac{7}{8}$ and $\frac{5}{8}$.*

$$\begin{array}{r} \frac{7}{8} \\ + \frac{5}{8} \\ \hline \frac{12}{8} \end{array} = \frac{3}{2} \text{ or } 1\frac{1}{2}$$

Simplify the sum.

The sum of $\frac{7}{8}$ and $\frac{5}{8}$ is $1\frac{1}{2}$.

To add fractions with unlike denominators, rename the fractions with a common denominator. Then add the fractions.

EXAMPLE *Find the sum of $\frac{1}{9}$ and $\frac{5}{6}$.*

$$\frac{1}{9} = \frac{2}{18}$$

Use 18 for the common denominator.

$$+\frac{5}{6} = \frac{15}{18}$$

$$\frac{17}{18}$$

The sum of $\frac{1}{9}$ and $\frac{5}{6}$ is $\frac{17}{18}$.

EXERCISES *Add.*

1. $\frac{4}{7} + \frac{2}{7}$

2. $\frac{5}{9} + \frac{4}{9}$

3. $\frac{11}{15} + \frac{2}{15}$

4. $\begin{array}{r} \frac{11}{15} \\ + \frac{7}{15} \\ \hline \end{array}$

5. $\begin{array}{r} \frac{6}{7} \\ + \frac{6}{7} \\ \hline \end{array}$

6. $\begin{array}{r} \frac{11}{12} \\ + \frac{5}{12} \\ \hline \end{array}$

7. $\begin{array}{r} \frac{1}{8} \\ + \frac{1}{9} \\ \hline \end{array}$

8. $\begin{array}{r} \frac{1}{3} \\ + \frac{1}{6} \\ \hline \end{array}$

9. $\begin{array}{r} \frac{3}{5} \\ + \frac{2}{7} \\ \hline \end{array}$

10. $\frac{7}{16} + \frac{3}{8}$

11. $\frac{7}{10} + \frac{2}{5}$

12. $\frac{3}{14} + \frac{1}{7}$

13. $\frac{5}{12} + \frac{1}{3}$

14. $\frac{1}{6} + \frac{1}{8}$

15. $\frac{1}{6} + \frac{4}{9}$

16. $\frac{3}{8} + \frac{5}{8} + \frac{1}{8}$

17. $\frac{1}{2} + \frac{1}{3} + \frac{1}{4}$

18. $\frac{2}{3} + \frac{3}{4} + \frac{1}{6}$

APPLICATIONS

19. After running $\frac{7}{8}$ mile in a horse race, a horse ran an additional $\frac{3}{8}$ mile to cool down. How far did the horse run altogether?

20. In 1991, about $\frac{1}{5}$ of the crude oil produced was from North America, and about $\frac{2}{7}$ of the crude oil produced was from the Middle East. What fraction of the crude oil produced was from North America or the Middle East?

21. In 1991, about $\frac{3}{10}$ of the petroleum consumed was in North America, and about $\frac{1}{5}$ of the petroleum consumed was in Western Europe. What fraction of the petroleum consumed was in North America or Western Europe?

SKILL 27

Name ____________________ Date __________

Subtracting Fractions

To subtract fractions with like denominators, subtract the numerators. Write the difference over the common denominator. Simplify the difference if possible.

EXAMPLE *Subtract $\frac{5}{12}$ from $\frac{7}{12}$.*

$$\begin{array}{r} \frac{7}{12} \\ -\frac{5}{12} \\ \hline \frac{2}{12} = \frac{1}{6} \end{array}$$

Simplify the difference.

The difference is $\frac{1}{6}$.

To subtract fractions with unlike denominators, rename the fractions with a common denominator. Then subtract the fractions.

EXAMPLE *Subtract $\frac{5}{8}$ from $\frac{5}{6}$.*

$$\begin{array}{rcr} \frac{5}{6} & = & \frac{20}{24} \\ -\frac{5}{8} & = & \frac{15}{24} \\ \hline & & \frac{5}{24} \end{array}$$

Use 24 for the common denominator.

The difference is $\frac{5}{24}$.

EXERCISES *Subtract.*

1. $\begin{array}{r} \frac{3}{4} \\ -\frac{1}{4} \\ \hline \end{array}$

2. $\begin{array}{r} \frac{5}{7} \\ -\frac{3}{7} \\ \hline \end{array}$

3. $\begin{array}{r} \frac{11}{12} \\ -\frac{3}{12} \\ \hline \end{array}$

4. $\dfrac{7}{16} - \dfrac{3}{16}$

5. $\dfrac{9}{10} - \dfrac{3}{10}$

6. $\dfrac{11}{12} - \dfrac{5}{12}$

7. $\dfrac{11}{12} - \dfrac{1}{3}$

8. $\dfrac{8}{15} - \dfrac{2}{5}$

9. $\dfrac{4}{5} - \dfrac{1}{10}$

10. $\frac{17}{18} - \frac{2}{9}$

11. $\frac{7}{8} - \frac{1}{3}$

12. $\frac{3}{4} - \frac{2}{5}$

13. $\frac{2}{5} - \frac{1}{6}$

14. $\frac{11}{12} - \frac{2}{3}$

15. $\frac{5}{6} - \frac{5}{8}$

16. $\frac{7}{12} - \frac{3}{10}$

17. $\frac{7}{9} - \frac{1}{6}$

18. $\frac{4}{7} - \frac{1}{2}$

19. $\frac{3}{4} - \frac{2}{5}$

20. $\frac{7}{8} - \frac{1}{3}$

21. $\frac{2}{3} - \frac{3}{5}$

APPLICATIONS

22. A large orange weighs $\frac{11}{16}$ pound. A small orange weighs $\frac{5}{16}$ pound. How much more does the large orange weigh?

23. In 1991, North America produced $\frac{1}{4}$ of the world's coal. The only area that produced more coal was the Far East, which produced $\frac{3}{8}$ of the coal. How much more of the world's coal was produced by the Far East than North America?

24. In 1991, North America consumed about $\frac{1}{5}$ of the coal produced and Western Europe consumed about $\frac{1}{7}$ of the coal produced. How much more coal was consumed by North America than Western Europe?

25. A page of a book has a $\frac{1}{2}$-inch margin on the top and a $\frac{3}{4}$-inch margin on the bottom. How much deeper is the bottom margin than the top margin?

SKILL 28 Name ______________________ Date __________

Multiplying Fractions

To multiply fractions, multiply the numerators. Then multiply the denominators. Simplify the product if possible.

EXAMPLES ***Multiply $\frac{4}{7}$ times $\frac{5}{9}$.***

$\frac{4}{7} \times \frac{5}{9} = \frac{4 \times 5}{7 \times 9}$ *Multiply the numerators. Multiply the denominators.*

$= \frac{20}{63}$

The product of $\frac{4}{7}$ and $\frac{5}{9}$ is $\frac{20}{63}$.

Multiply $\frac{5}{6}$ times $\frac{3}{5}$.

$\frac{5}{6} \times \frac{3}{5} = \frac{5 \times 3}{6 \times 5}$ *Multiply the numerators. Multiply the denominators.*

$= \frac{15}{30}$ or $\frac{1}{2}$ *Simplify.*

The product of $\frac{5}{6}$ and $\frac{3}{5}$ is $\frac{1}{2}$.

EXERCISES ***Multiply.***

1. $\frac{2}{3} \times \frac{1}{4}$

2. $\frac{3}{7} \times \frac{1}{2}$

3. $\frac{1}{3} \times \frac{3}{5}$

4. $\frac{1}{2} \times \frac{6}{7}$

5. $\frac{7}{10} \times \frac{5}{7}$

6. $\frac{1}{4} \times \frac{1}{4}$

7. $\frac{1}{3} \times \frac{1}{5}$

8. $\frac{5}{8} \times \frac{1}{2}$

9. $\frac{4}{9} \times \frac{3}{4}$

10. $\frac{2}{3} \times \frac{3}{8}$

11. $\frac{1}{7} \times \frac{1}{7}$

12. $\frac{2}{9} \times \frac{1}{2}$

13. $\frac{3}{5} \times \frac{5}{6}$

14. $\frac{2}{7} \times \frac{1}{3}$

15. $\frac{5}{12} \times \frac{1}{5}$

16. $\frac{1}{2} \times \frac{1}{5}$

17. $\frac{6}{7} \times \frac{8}{15}$

18. $\frac{8}{9} \times \frac{9}{10}$

19. $\frac{4}{5} \times \frac{5}{14}$

20. $\frac{7}{8} \times \frac{4}{9}$

21. $\frac{5}{8} \times \frac{3}{4}$

APPLICATIONS ***Use the recipe for lemon chicken saute below to answer Exercises 22–25.***

6 boneless chicken breasts, rolled in flour	$\frac{1}{3}$ cup teriyaki sauce
$\frac{1}{4}$ cup butter	$\frac{1}{2}$ teaspoon sugar
3 tablespoons lemon juice	$\frac{1}{8}$ teaspoon pepper
1 teaspoon garlic	

22. If Julie wants to make half of this recipe, how much pepper should she use?

23. If Julie wants to make one-third of this recipe, how much teriyaki sauce should she use?

24. If Julie wants to make two-thirds of this recipe, how much sugar should she use?

25. If Julie wants to make two-thirds of this recipe, how much butter should she use?

26. If about $\frac{1}{3}$ of Earth is able to be farmed and $\frac{2}{5}$ of this land is planted in grain crops, what part of Earth is planted in grain crops?

27. Two fifths of the students at Main Street Middle School are in seventh grade. If half of the students in seventh grade are boys, what fraction of the students are seventh grade boys?

Name ______________________ Date __________

Multiplying Whole Numbers by Fractions

Mr. Quin's class has 28 students. He has enough computers for $\frac{3}{4}$ of the class members to work on the computers at any given time.

EXAMPLE ***How many students can use the computers at a time?***

$$\frac{3}{4} \times 28 = \frac{3}{4} \times \frac{28}{1}$$
$$= \frac{84}{4}$$
$$= 21$$

Twenty-one students can use the computers at a time.

EXERCISES ***Multiply. Write each product in simplest form.***

1. $\frac{1}{2} \times 50$
2. $\frac{1}{5} \times 30$
3. $\frac{1}{3} \times 6$
4. $\frac{1}{2} \times 9$
5. $7 \times \frac{1}{7}$
6. $15 \times \frac{1}{5}$
7. $\frac{2}{3} \times 9$
8. $\frac{3}{5} \times 10$
9. $16 \times \frac{3}{4}$

10. $\frac{5}{6} \times 8$

11. $14 \times \frac{1}{3}$

12. $24 \times \frac{7}{6}$

13. $\frac{4}{5} \times 25$

14. $18 \times \frac{1}{5}$

15. $16 \times \frac{3}{2}$

16. $20 \times \frac{5}{4}$

17. $\frac{1}{2} \times 11$

18. $\frac{5}{7} \times 28$

APPLICATIONS ***A Native-American recipe for hickory nut corn pudding is given at the right. Use the recipe to answer Exercises 19–21.***

Hickory Nut Corn Pudding

$1\frac{1}{2}$ cups cooked corn

$\frac{1}{2}$ cup shelled dried hickory nuts, chopped

2 tablespoons nut butter

1 cup boiling water

2 eggs, beaten

2 tablespoons honey

2 tablespoons corn meal

$\frac{1}{4}$ cup raisins

Combine all ingredients into a well-greased casserole dish. Bake at 350°F for 1 hour. Serve hot.

19. How many cups of hickory nuts should be used if the recipe is tripled?

20. How many cups of raisins should be used if the recipe is to be multiplied by 6?

21. How much corn meal should be used if the recipe is to be cut by one third?

22. A meteorologist in a midwestern city checked the weather records for the first 90 days of the year for the past several years. She observed that each year about $\frac{2}{3}$ of these days were sunny. How many of the first 90 days of this coming year should she expect to be sunny?

23. In 1936, Franklin D. Roosevelt won the presidential election with about $\frac{3}{5}$ of the popular vote. There were about 46,000,000 votes cast in that election. About how many popular votes did F.D.R. receive?

24. About $\frac{1}{3}$ of the people living in Africa live in urban areas. In 1992, there were about 681,700,000 people living in Africa. About how many people lived in urban areas?

SKILL 30

Name ______________________ Date ________

Dividing Fractions

To divide by a fraction, multiply by its reciprocal. Simplify the quotient if possible.

EXAMPLES ***Divide $\frac{2}{3}$ by $\frac{5}{7}$.***

$\frac{2}{3} \div \frac{5}{7} = \frac{2}{3} \times \frac{7}{5}$ *Multiply by the reciprocal of $\frac{5}{7}$.*

$= \frac{2 \times 7}{3 \times 5}$ *Multiply the numerators.* *Multiply the denominators.*

$= \frac{14}{15}$

The quotient is $\frac{14}{15}$.

Divide $\frac{3}{4}$ by $\frac{9}{10}$.

$\frac{3}{4} \div \frac{9}{10} = \frac{3}{4} \times \frac{10}{9}$ *Multiply by the reciprocal of $\frac{9}{10}$.*

$= \frac{3 \times 10}{4 \times 9}$ *Multiply the numerators.* *Multiply the denominators.*

$= \frac{30}{36}$ or $\frac{5}{6}$ *Simplify.*

The quotient is $\frac{5}{6}$.

EXERCISES ***Divide.***

1. $\frac{3}{4} \div \frac{1}{2}$

2. $\frac{4}{5} \div \frac{1}{3}$

3. $\frac{1}{5} \div \frac{1}{4}$

4. $\frac{4}{7} \div \frac{8}{9}$

5. $\frac{3}{8} \div \frac{3}{4}$

6. $\frac{9}{7} \div \frac{3}{14}$

7. $\frac{4}{5} \div \frac{2}{5}$

8. $\frac{7}{8} \div \frac{1}{4}$

9. $\frac{2}{5} \div \frac{5}{8}$

10. $\frac{1}{3} \div \frac{1}{6}$

11. $\frac{5}{8} \div \frac{5}{12}$

12. $\frac{4}{5} \div \frac{2}{7}$

13. $\frac{2}{5} \div \frac{3}{10}$

14. $\frac{5}{7} \div \frac{3}{4}$

15. $\frac{2}{3} \div \frac{4}{9}$

16. $\frac{4}{7} \div \frac{4}{5}$

17. $\frac{5}{6} \div \frac{1}{9}$

18. $\frac{4}{5} \div \frac{2}{3}$

APPLICATIONS

19. About $\frac{1}{20}$ of the population of the world lives in South America. If about $\frac{1}{35}$ of the population of the world lives in Brazil, what fraction of the population of South America lives in Brazil?

20. Three fourths of a pizza is left. If the pizza was originally cut in $\frac{1}{8}$ pieces, how many pieces are left?

The area of each rectangle is given. Find the missing length for each rectangle.

21.

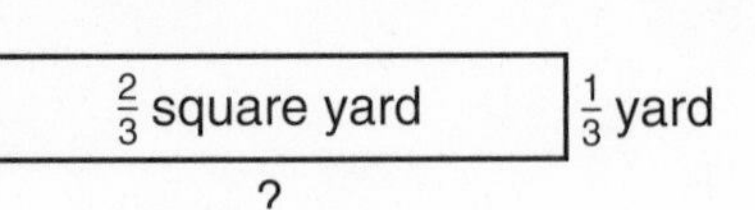

22.

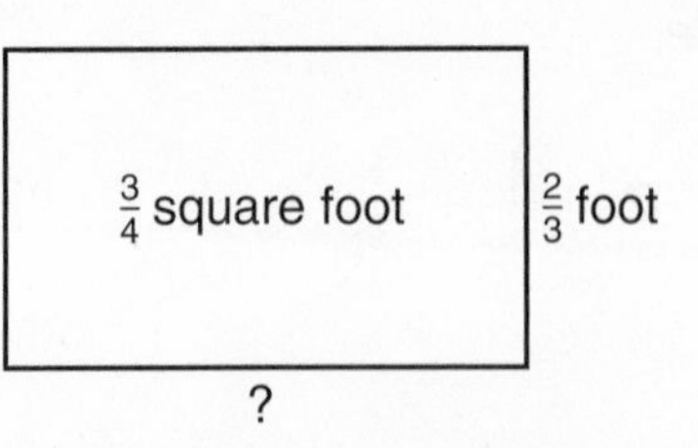

23.

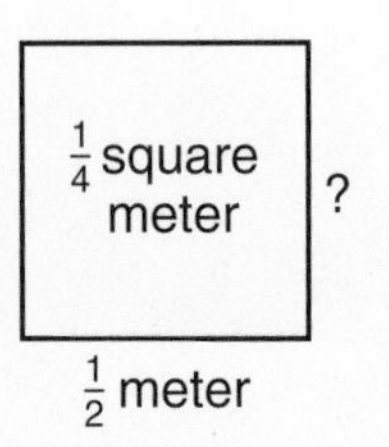

24.

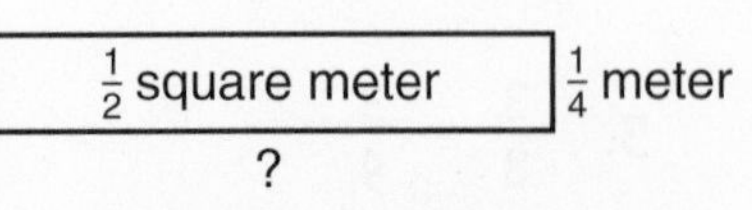

SKILL 31

Name ______________________ Date __________

Adding Integers

You can also add integers on a number line. Locate the first addend on the number line. Move right if the second addend is positive. Move left if the second addend is negative.

EXAMPLE ***Find 4 + (–10).***

Start at 4. Since –10 is negative, move left 10 units.

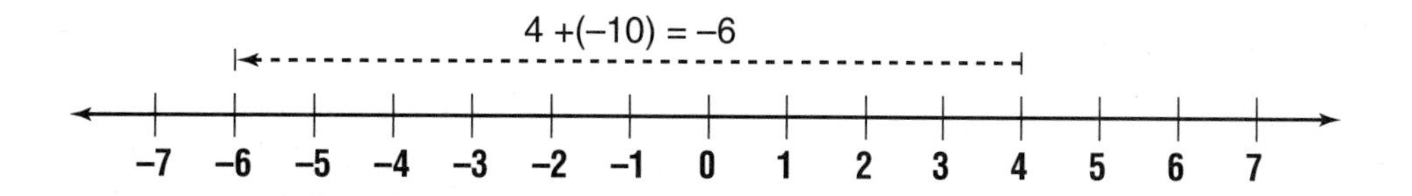

Therefore, 4 + (–10) = –6.

When you add integers, remember:

- The sum of two positive integers is positive.
- The sum of two negative integers is negative.
- The sum of a positive and a negative integer is:
  positive if the positive integer has the greater absolute value.
  negative if the negative integer has the greater absolute value.

EXERCISES ***State whether each sum is positive or negative.***

1. –4 + (–2)

2. 5 + (–3)

3. –10 + 7

4. 9 + (–3)

5. 6 + 0

6. –8 + (–1)

Find each sum. Use counters or a number line if necessary.

7. $3 + (-6)$	**8.** $-9 + 8$	**9.** $-4 + 7$
10. $6 + (-6)$	**11.** $-8 + (-2)$	**12.** $2 + (-5)$
13. $9 + (-18)$	**14.** $-8 + (-7)$	**15.** $14 + (-6)$
16. $-12 + 5$	**17.** $-4 + 10$	**18.** $-9 + (-8)$
19. $3 + (-11)$	**20.** $-6 + 13$	**21.** $-12 + 6$
22. $-22 + (-7)$	**23.** $-6 + 36$	**24.** $-38 + 6$

APPLICATIONS

25. At 5:00 P.M., Sandra Jean read a temperature of 42°F. What was the temperature after it changed –6 degrees?

26. A scuba diver descends to a depth of 30 meters below sea level and then rises 23 meters to study a school of fish. How far below the surface is the school of fish?

27. Jack's score at the end of a game was 45. Halfway through the game his score was –10. How many points did he score during the last half of the game?

SKILL 32

Name ______________________________ Date __________

Subtracting Integers

An integer and its opposite are the same distance from 0 on a number line. The integers 5 and –5 are opposites.

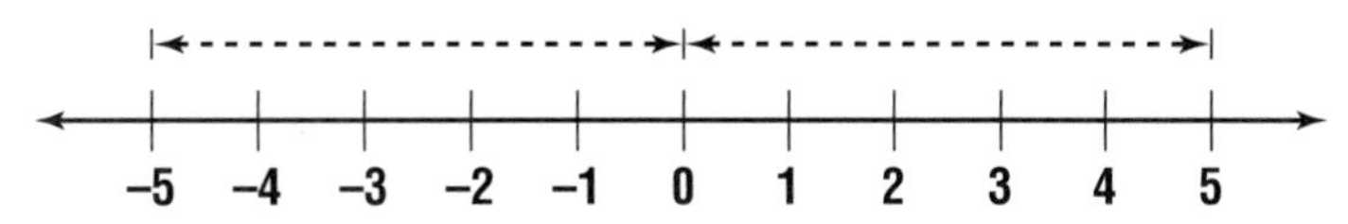

To subtract an integer, add its opposite.

EXAMPLES ***Find 6 – 9.***

$6 - 9 = 6 + (-9)$ *To subtract 9, add –9.*

$= -3$

Find –10 – (–12).

$-10 - (-12) = -10 + 12$ *To subtract –12, add 12.*

$= 2$

EXERCISES ***Find the opposite of each integer.***

1. –18 **2.** 12 **3.** –36 **4.** 61

Rewrite each subtraction problem as an addition problem.

5. 5 – (–3) **6.** –6 – (–1) **7.** –8 – 7

8. –2 – 10 **9.** 15 – (–15) **10.** 6 – 11

Find each difference. Use counters if necessary.

11. $-8 - (-4)$ **12.** $7 - (-5)$ **13.** $-4 - 2$

14. $-3 - (-5)$ **15.** $6 - (-10)$ **16.** $8 - 5$

17. $-1 - 4$ **18.** $2 - (-2)$ **19.** $-5 - (-1)$

20. $7 - 2$ **21.** $-9 - (-6)$ **22.** $8 - (-2)$

23. $-8 - (-14)$ **24.** $13 - (-7)$ **25.** $19 - (-6)$

26. $-17 - 13$ **27.** $-11 - (-20)$ **28.** $-10 - (-10)$

APPLICATIONS

29. The elevation of Denver, Colorado, is 5,280 feet above sea level. In contrast, New Orleans, Louisiana, has an elevation of 5 feet below sea level. What is the difference in their elevations?

30. The monthly profit for a small business is found by subtracting the expenses from the income. In a certain month, a business had expenses of $2,500 and income of $2,000. What was the profit for that month?

31. The sum of –5 and a certain number is 10. What is that number?

32. At night, the average temperature on the surface of the planet Saturn is –150°C. During the day, the average temperature is –123°C. What is the difference in temperature?

SKILL 33

Name ______________________ Date __________

Adding and Subtracting Integers

You can use the following rules when adding or subtracting integers.

- The sum of two positive integers is positive.
- The sum of two negative integers is negative.
- To add integers with different signs, subtract their absolute value. The sum is positive if the positive integer has the greater absolute value. The sum is negative if the negative integer has the greater absolute value.
- To subtract an integer, add its opposite.

EXAMPLE ***Frank withdrew $45 from his checking account. Then he withdrew $15.95 more. What was the change in Frank's balance?***

A withdrawal of $45 can be indicated by –45 and a withdrawal of $15.95 can be indicated by –15.95. To find the change, add –45 and –15.95.

$$-45 + (-15.95) = -60.95$$

The change in Frank's balance was –$60.95.

EXERCISES ***Add or subtract.***

1. $-14 + (-2)$ **2.** $6 + (-5)$ **3.** $-2 + 17$

4. $-18 + (-3)$ **5.** $-27 + 1$ **6.** $-20 + 35$

7. $15 - (-20)$ **8.** $-5 + (-17)$ **9.** $-7 + (-7)$

10. $-11 - (-8)$ **11.** $-14 + 13$ **12.** $-48 - 48$

13. $-16 + 24$ **14.** $-50 - 3$ **15.** $-2 - (-34)$

16. $-48 + 21$ **17.** $44 + (-4)$ **18.** $-75 - 6$

19. $-15 - 7$ **20.** $-7 + (-31)$ **21.** $65 + (-10)$

22. $-22 - 17$

23. $32 + (-16)$

24. $-41 - (-9)$

25. $-102 + (-5)$

26. $33 + (-43)$

27. $100 - (-20)$

28. $-81 - 9$

29. $28 + (-56)$

30. $-14 - (-43)$

APPLICATIONS ***A submarine is 1,200 meters below sea level. Use this information to answer Exercises 31–33.***

31. The submarine descends an additional 950 meters. How far below sea level is the submarine now?

32. The submarine then ascends 700 meters. How far below sea level is the submarine now?

33. The submarine ascends an additional 1,050 meters. How much more does it need to ascend to be at sea level?

34. The highest point in New Orleans, Louisiana is 15 feet above sea level. The lowest point is 4 feet below sea level. How much higher is the highest point than the lowest point?

35. Mark Little's checking account is overdrawn by $35. How much does he need to deposit to have a balance of $100?

36. The temperature outside was 15°F. The wind chill made it feel like –20°F. What is the difference between these two temperatures?

SKILL 34

Name ______________________ Date __________

Multiplying Integers

When multiplying integers:

The product of two positive integers is positive.

The product of two negative integers is positive.

The product of a positive integer and a negative integer is negative.

EXAMPLE ***If the average daily temperature changed –2°F each day for a week, how much did the temperature change in all?***

There are 7 days in a week.

You can multiply 7 times –2°F and determine the week's temperature change.

Solve: $n = 7(-2)$

$n = -14$

The temperature changed –14°F during the week.

EXERCISES ***Solve each equation.***

1. $-7(-8) = p$ **2.** $10(-6) = j$ **3.** $a = -9(3)$

4. $(-8)^2 = e$ **5.** $m = (-12)(-12)$ **6.** $t = (-25)(4)$

7. $4(-12)(-3) = r$ **8.** $h = -2(24)$ **9.** $k = (-4)(35)(-1)$

Evaluate each expression if m = –6, n = 3, and p = –4.

10. np **11.** $-5np$ **12.** $2mn$

13. $-12np$ **14.** $-10mp$ **15.** $-2m^2$

Simplify each of the following.

16. $-8(-3) + (-3)8$

17. $10(-4) - (-4)(-10)$

18. Which product is *not* equal to 16?

a. $-4(-4)$ b. $8(-2)$ c. $-16(-1)$ d. $-8(-2)$

19. Which product is *not* equal to 20?

a. $-10(-2)$ b. $-5(-4)$ c. $(-20)(-1)$ d. $5(-4)$

For each set of numbers, identify the pattern. Then write the next three numbers in the pattern.

20. 1, –2, 4, –8, 16, . . .

21. –2, –10, –50, –250, –1,250, . . .

22. –1, 3, –9, 27, –81, . . .

APPLICATION

23. Find your way through the maze by moving to the expression with the next highest value.

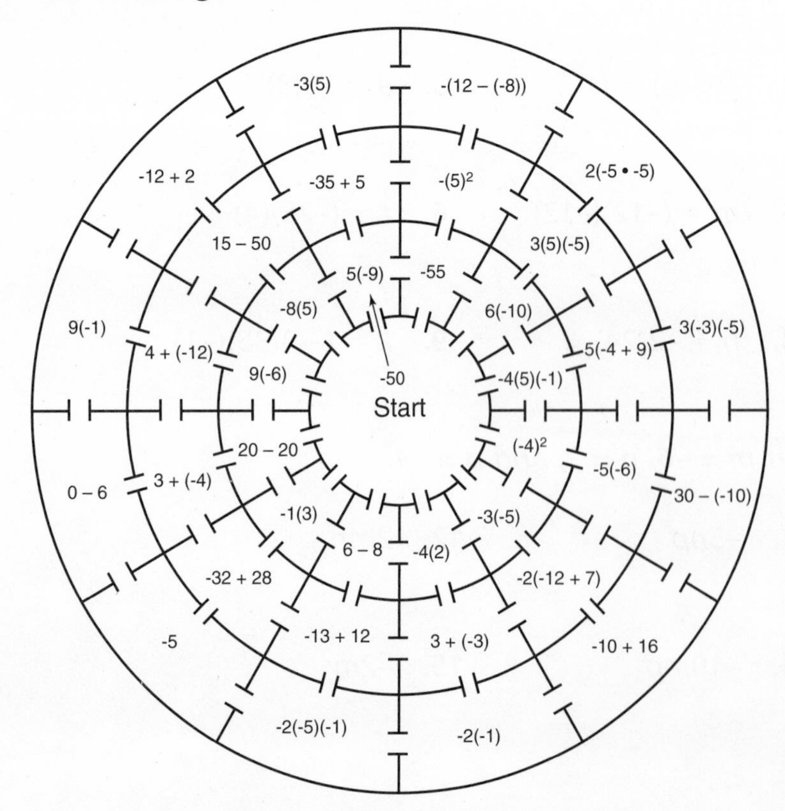

SKILL 35

Name ______________________ Date ____________

Dividing Integers

If two integers have the same sign, their quotient is positive.

EXAMPLES ***Solve each equation.***

$k = 560 \div 8$ *Both integers are positive.*
$k = 70$ *The quotient is positive.*

The solution is 70.

$h = -120 \div (-6)$ *Both integers are negative.*
$h = 20$ *The quotient is positive.*

The solution is 20.

If two integers have different signs, their quotient is negative.

EXAMPLES ***Solve each equation.***

$a = -75 \div 5$ *The dividend is negative, the divisor is positive.*
$a = -15$ *The quotient is negative.*

The solution is –15.

$b = 99 \div (-33)$ *The dividend is positive, the divisor is negative.*
$b = -3$ *The quotient is negative.*

The solution is –3.

EXERCISES ***Solve each equation.***

1. $f = -16 \div (-4)$ **2.** $m = -28 \div 7$ **3.** $52 \div (-4) = g$

4. $-32 \div (-16) = q$ **5.** $d = -200 \div (-25)$ **6.** $84 \div (-6) = h$

7. $-51 \div 3 = n$ **8.** $256 \div (-32) = k$ **9.** $-132 \div (-11) = y$

10. $x = -75 \div (-15)$ **11.** $p = -68 \div 4$ **12.** $z = -116 \div (-29)$

Evaluate each expression if e = –36, f = 4, and g = –3.

13. $eg \div f$

14. $-48 \div g$

15. $e^2 \div f$

16. $\frac{e}{f}$

17. $\frac{e}{fg}$

18. $\frac{e}{g^2}$

19. $\frac{e^2}{fg}$

20. $\frac{-100}{f}$

21. $\frac{e^2}{g^2}$

22. Which quotient equals 7?
a. $-21 \div 3$ **b.** $14 \div (-2)$ **c.** $28 \div (-4)$ **d.** $-21 \div (-3)$

23. Which quotient is *not* equal to –4?
a. $12 \div (-3)$ **b.** $-24 \div (-6)$ **c.** $-20 \div 5$ **d.** $-4 \div 1$

24. Which quotient is *not* equal to 5?
a. $15 \div 3$ **b.** $-20 \div (-4)$ **c.** $-15 \div 3$ **d.** $-25 \div (-5)$

APPLICATIONS

25. Using the formula $F = \frac{9}{5}C + 32$, find the Fahrenheit temperature for a Celsius temperature of –5 degrees.

26. Amanda kept track of the 7:00 A.M. temperature every day for a week. The following is her chart in degrees Fahrenheit. What was the average temperature for the week?

Sun.	Mon.	Tues.	Wed.	Thurs.	Fri.	Sat.
–11	–9	–7	–6	0	2	3

27. The San Francisco 49ers had trouble with penalties during a Monday night game. They lost a total of 35 yards. Two of the penalties were for 10 yards each, and the other three were each for the same number of yards. How many yards were each of the other three penalties?

Name ______________________________ Date __________

Multiplying and Dividing Integers

You can use the following rules when multiplying or dividing integers.

- The product of two integers with the same sign is positive.
- The product of two integers with different signs is negative.
- The quotient of two integers with the same sign is positive.
- The quotient of two integers with different signs is negative.

EXAMPLE ***Sabrina is thinking of two pairs of integers whose product is –88 and whose quotient is –22. What are the integers?***

Guess and check to find the integers.

Guess 1: Try –11 and 8.
$-11 \cdot 8 = -88$
$-11 \div 8 \neq -22$

Guess 2: Try –22 and 4.
$-22 \cdot 4 = -88$
$-22 \div 4 \neq -22$

Guess 3: Try –44 and 2.
$-44 \cdot 2 = -88$
$-44 \div 2 = -22$

One pair of integers is –44 and 2. The other pair of integers is 44 and –2 since $44 \cdot -2 = -88$ and $44 \div -2 = -22$.

EXERCISES ***Multiply or divide.***

1. $-4 \cdot (-2)$ **2.** $-5 \cdot 3$ **3.** $9 \cdot (-4)$ **4.** $-7 \cdot (-4)$

5. $-36 \div (-6)$ **6.** $-48 \div 6$ **7.** $72 \div (-8)$ **8.** $-28 \div 4$

9. $-3 \cdot 20$ 10. $-40 \div (-2)$ 11. $25 \cdot (-3)$ 12. $80 \div (-20)$

13. $-70 \div (-2)$ 14. $-50 \cdot 3$ 15. $-33 \cdot (-2)$ 16. $12 \cdot (-4)$

17. $44 \div (-4)$ 18. $-75 \div 3$ 19. $-15 \cdot (-5)$ 20. $-16 \cdot (-3)$

21. $-125 \div (-25)$ 22. $200 \div (-100)$ 23. $-20 \cdot (-100)$ 24. $400 \cdot (-6)$

25. $500 \div (-5)$ 26. $-110 \cdot 8$ 27. $-900 \div (-45)$ 28. $-560 \div 7$

29. $80 \cdot (-30)$ 30. $-60 \cdot 12$ 31. $-450 \div (-90)$ 32. $-25 \cdot (-40)$

APPLICATIONS ***Suppose the temperature has been changing at a rate of –3°F per hour and the temperature is predicted to continue changing at this rate. The current temperature is 5°F. Use this information to answer Exercises 33–35.***

33. What was the temperature 2 hours ago?

34. What is the temperature predicted to be 3 hours from now?

35. In how many hours is the temperature predicted to be –19°F?

36. The lowest temperature ever recorded in Alaska is –62°C. Use the formula $C = \frac{9}{5}F + 32$ to find this temperature in degrees Fahrenheit. Round your answer to the nearest degree.

37. Marta is thinking of two pairs of integers whose product is –400 and whose quotient is –25. What are the integers?

38. Ned is thinking of two pairs of integers whose product is 40 and whose quotient is 10. What are the integers?

SKILL 37

Name ______________________ Date __________

Solve Equations Involving Addition

To solve an equation means to find a value for the variable that makes the equation true. To solve an equation, you need to get the variable by itself.

Subtraction Property of Equality: If you subtract the same number from each side of an equation, the two sides remain equal.

EXAMPLE ***Solve $t + 12.2 = 25.1$.***

$$t + 12.2 = 25.1$$

$$t + 12.2 - 12.2 = 25.1 - 12.2 \quad \textit{Subtract 12.2 from each side.}$$

$$t = 12.9$$

Check:

$$t + 12.2 = 25.1$$

$$12.9 + 12.2 \stackrel{?}{=} 25.1 \quad \textit{Replace t with 12.9.}$$

$$25.1 = 25.1 \checkmark$$

The solution is 12.9.

EXERCISES ***Solve each equation. Check your solution.***

1. $b + 7 = 22$

2. $r + 0.4 = 11.5$

3. $45 = t + 17$

4. $17 + k = 62$

5. $146 + j = 199$

6. $17.2 = h + 4.9$

7. $n + 2\frac{1}{3} = 4\frac{2}{3}$

8. $5\frac{2}{5} + v = 7\frac{1}{2}$

9. $x + 7\frac{1}{2} = 20$

10. $18.42 + t = 63$

11. $e + 12.2 = 40$

12. $m + 18 = 78$

APPLICATIONS

13. Cicely is saving money to buy a computer printer that costs $399. She has already saved $150. If y stands for the amount she still needs to save, which equation could you solve to find the amount she still needs to save?

- **a.** $150 + 399 = y$
- **b.** $399 + y = 150$
- **c.** $150 + y = 399$
- **d.** none of these

14. The George Washington Carver National Monument is 263 acres smaller than the 473-acre Casa Grande National Monument. Solve the equation $g + 263 = 473$ to find the size of the George Washington Carver National Monument.

15. Wayne bought a share of stock at $29\frac{3}{4}$. A year later, the stock was selling for $42\frac{1}{8}$. How much would Wayne have gained if he had sold his stock then?

16. Jamal delivers 60 papers each day after school. Today he has already delivered 22 papers. Find how many more papers he must deliver by writing an equation and solving it.

17. For Jane's girl scout troop, she needs to volunteer a total of 150 hours in order to earn her Community Service patch. She has volunteered 67 hours already. Find how many more hours she must volunteer by writing an equation and solving it.

18. There are 28 students in art class. Seven students in the class wear glasses or contact lenses. How many students do *not* wear glasses or contact lenses?

Name ____________________ Date __________

Solve Equations Involving Subtraction

To solve an equation means to find a value for the variable that makes the equation true. To solve an equation, you need to get the variable by itself.

Addition Property of Equality: If you add the same number to each side of an equation, the two sides remain equal.

EXAMPLE ***Solve $t - 12.2 = 25.1$.***

$$t - 12.2 = 25.1$$

$$t - 12.2 + 12.2 = 25.1 + 12.2 \quad \text{Add 12.2 to each side.}$$

$$t = 37.3$$

Check:

$$t - 12.2 = 25.1$$

$$37.3 - 12.2 \stackrel{?}{=} 25.1 \quad \text{Replace } t \text{ with 37.3.}$$

$$25.1 = 25.1 \ \checkmark$$

The solution is 37.3.

EXERCISES ***Solve each equation. Check your solution.***

1. $8.9 = p - 3.3$

2. $j - 4.5 = 1.7$

3. $y - 9 = 29$

4. $p - 23\frac{4}{5} = 35\frac{7}{10}$

5. $w - 6\frac{1}{2} = 18$

6. $f - 19 = 77$

7. $m - 9.4 = 15.7$

8. $153 = k - 23$

9. $u - 27 = 12$

10. $p - 58 = 73$

11. $x - 4.9 = 12.2$

12. $105 = y - 17$

APPLICATIONS

13. Madaline was filling balloons with helium for a party. She filled 24 balloons. While she was filling those, she filled 7 others too full and they burst. If t stands for the total number of ballons that she filled, which equation could you solve to find the total number of balloons that she filled?

a. $24 - 7 = t$
b. $24 - t = 7$
c. $t - 7 = 24$
d. none of these

14. Joe and José have a painting business. Joe spent 3.75 hours painting three rooms of the Dutton's house. This was 6.75 hours less than the total time it took to do the job. Find how much time it took to paint the three rooms by writing an equation and solving it.

15. Ryan and Nick went to the fair. When they rode the carousel, Ryan counted 10 horses that were stationary. This was 24 less than the total number of horses on the carousel. Find how many total horses were on the carousel by writing an equation and solving it.

16. Ben has an insect and spider collection. Fifteen of the bugs are spiders. This is 8 less than the total number of bugs that he has. Find how many bugs Ben has in his collection by writing an equation and solving it.

17. Pat spent $575 buying blankets for the homeless shelter. Cash Mart said that they would match the number of blankets that all of Pat's friends brought to the shelter. Pat's friends brought 23 blankets, but Cash Mart actually gave 45 blankets. A total of 100 blankets were donated. How many blankets were donated by people other than Cash Mart and Pat's friends?

Name ______________________ Date __________

Solve Equations Involving Multiplication

You can use equations to solve multiplication problems. When a variable is multiplied by a number, divide each side of the equation by that number to set the variable by itself.

Division Property of Equality: If you divide each side of an equation by the same nonzero number, the two sides remain equal.

EXAMPLE ***Solve $48.6 = 6c$.***

$$48.6 = 6c$$

$$\frac{48.6}{6} = \frac{6c}{6} \qquad \textit{Divide each side by 6.}$$

$$8.1 = c$$

Check:

$$48.6 = 6c$$

$$48.6 \stackrel{?}{=} 6 \times 8.1 \qquad \textit{Replace } c \textit{ with 8.1.}$$

$$48.6 = 48.6 \checkmark$$

The solution is 8.1.

EXERCISES ***Solve each equation. Check your solution.***

1. $5r = 45$ **2.** $180 = 9v$ **3.** $17v = 289$

4. $5.1p = 61.2$ **5.** $6.4t = 64$ **6.** $91 = 13k$

7. $2.4(1.8) = w$ **8.** $\$8.46h = \$54.99$ **9.** $504 = 2.8m$

10. $9n = -45$ **11.** $5m = -35$ **12.** $-72 = 6r$

Write an equation for each of the following.

13. A bingo prize of $125 had to be split evenly among five people. How much did each person receive?

14. There are twice as many dogs as there are cats on Sheila's street. If there are six dogs, how many cats are there?

15. Each household on Tremont Street has two cameras. There are 370 cameras on this street. How many houses are there?

16. One hundred fifty six students in Barrington School own a moped. This is four times as many students as owned one three years ago. How many students owned one three years ago?

APPLICATIONS

17. The sum of the measures of the interior angles of a pentagon is 540°. The five angles all have the same measure. Solve the equation $5x = 540$ to find the measure of each angle.

18. At one gas station, one fourth of the customers buy premium gasoline. In one hour, 12 customers bought premium gasoline. What was the total number of customers for the hour?

19. Manuel's weekly pay check is $450. What is his annual salary?

20. A triangle has a base of 6 feet and an area of 9 square feet. What is its height? Remember the area of a triangle is half the base times the height.

21. What is the length of a rectangle with an area of 40 square feet and a width of 5 feet?

Name ______________________ Date __________

Solve Equations Involving Division

You can use equations to solve division problems. When a variable is divided by a number, multiply each side of the equation by that number to get the variable by itself.

Multiplication Property of Equality: If you multiply each side of an equation by the same number, the two sides remain equal.

EXAMPLE ***Solve*** $\frac{w}{5} = 2.3$.

$$\frac{w}{5} = 2.3$$

$$\frac{w}{5} \times 5 = 2.3 \times 5 \quad \text{Multiply each side by 5.}$$

$$w = 11.5$$

Check:

$$\frac{w}{5} = 2.3$$

$$\frac{11.5}{5} \stackrel{?}{=} 2.3 \quad \text{Replace } w \text{ with 11.5.}$$

$$2.3 = 2.3 \checkmark$$

The solution is 11.5.

EXERCISES ***Solve each equation. Check your solution.***

1. $\frac{1}{2}y = 7$

2. $30 = \frac{1}{5}k$

3. $\frac{x}{7} = 3$

4. $24 = \frac{r}{2.5}$

5. $\frac{w}{6} = 0.6$

6. $\frac{1}{3}c = \frac{3}{4}$

7. $\frac{y}{7} = 3.5$

8. $\frac{f}{1.1} = 7$

9. $\frac{1}{2} = \frac{1}{8}c$

10. $3.5 = \frac{m}{4}$

11. $\frac{y}{5} = 2.4$

12. $\frac{s}{8} = 9.6$

13. Solve $z \div \frac{1}{3} = \frac{6}{7}$.

a. $\frac{6}{21}$ **b.** $\frac{18}{7}$ **c.** $\frac{11}{21}$ **d.** $\frac{2}{7}$

14. Solve $w \div \frac{1}{5} = \frac{3}{4}$.

a. $\frac{1}{2}$ **b.** $\frac{15}{4}$ **c.** $\frac{3}{20}$ **d.** $\frac{2}{12}$

15. The quotient when the number e is divided by 18 is 8. Find the number.

16. The quotient when the number x is divided by 19 is 7. Find the number.

APPLICATIONS

17. Hisako can stay in the sun for 0.5 hours without burning. If she uses NEW LONGER TAN, that has a sun-protection factor of 30, she can safely bask in the sun three times as long. Use the equation $m \div 0.5 = 3$ to determine the number of hours Hisako can stay in the sun using NEW LONGER TAN.

18. One-half of the students who participated in the Walk-a-Thon got a T-shirt. If 28 T-shirts were given out, how many students participated in the Walk-a-Thon?

Name ________________________ Date ____________

The Coordinate System

The **coordinate system** is used to graph points in a plane. The horizontal line is the ***x*-axis.** The vertical line is the ***y*-axis.** Their intersection is called the **origin.**

Points are located using **ordered pairs.** The first number in an ordered pair is the *x*-coordinate; the second number is the *y*-coordinate.

EXAMPLES ***Name the ordered pair for point P.***

Start at the origin.
Move 3 units to the left along the *x*-axis.
Move 2 units up on the *y*-axis.
The ordered pair for point *P* is (–3, 2).

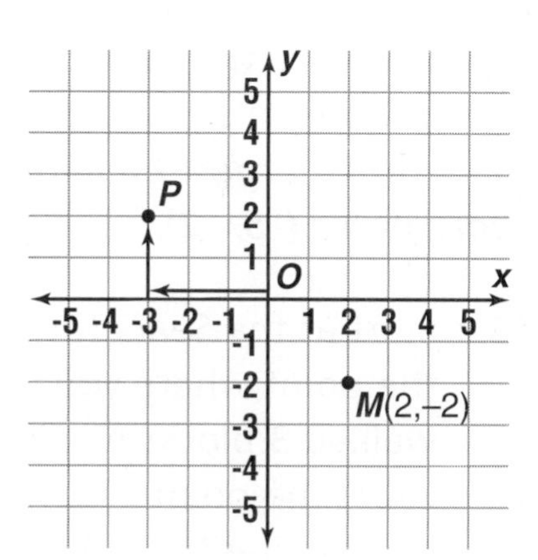

Graph the point M(2, –2).

Start at the origin.
Move 2 units to the right along the *x*-axis.
Move 2 units down on the *y*-axis.
Draw a point and label it *M*.

EXERCISES ***Name the ordered pair for each point on the coordinate plane.***

1. *A*
2. *B*
3. *C*
4. *D*
5. *E*
6. *F*

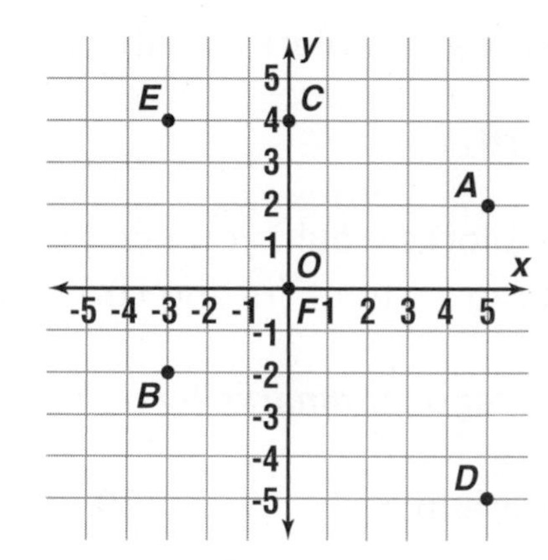

Graph and label each point on the coordinate plane.

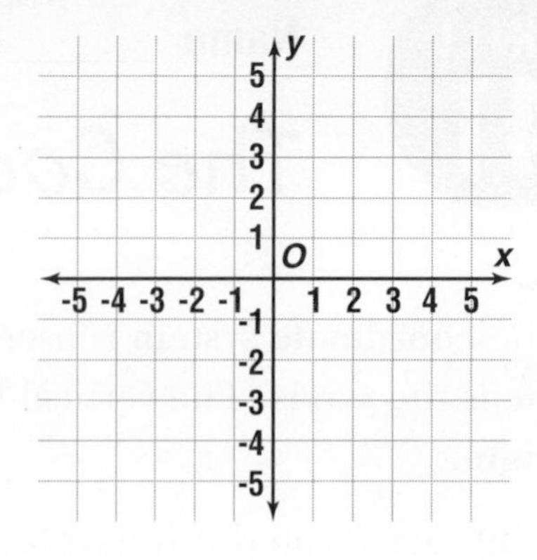

7. $N(-1, 3)$ **8.** $V(2, -4)$

9. $M(-2, 0)$ **10.** $K(-1, 5)$

11. $A(5, -1)$ **12.** $T(-3, 3)$

APPLICATIONS

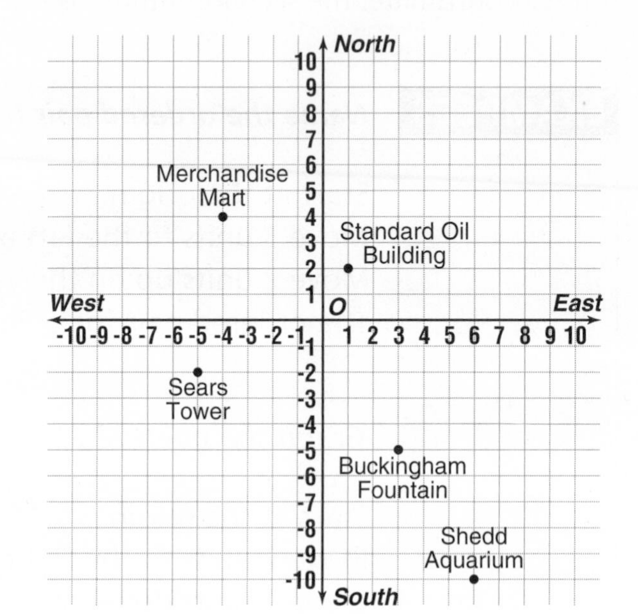

Chicago was planned in such a way that a rectangular coordinate system can describe locations very well. If a block is the distance between major streets, then the Sears Tower, the world's largest building, is located at the coordinates (–5, –2).

13. Start at the Sears Tower and graph the point where you would be if you walked 3 blocks north and 1 block east. Label the point "13." What are its coordinates?

For Exercises 14–18, start at point 13. Label each point and name its coordinates.

14. 4 blocks east and 2 blocks south

15. 2 blocks north and 1 block east

16. 7 blocks west and 6 blocks north

17. 3 blocks north

18. How many blocks would you need to walk to get back to the Sears Building without cutting corners?

Locate these Chicago landmarks by naming their coordinates.

19. Shedd Aquarium **20.** Buckingham Fountain

21. Standard Oil Building **22.** Merchandise Mart

Name ______________________ Date ________

Classifying Angles

An **angle** is formed by two rays with a common endpoint called the **vertex**. Angles are measured in degrees. Angles are classified according their measure.

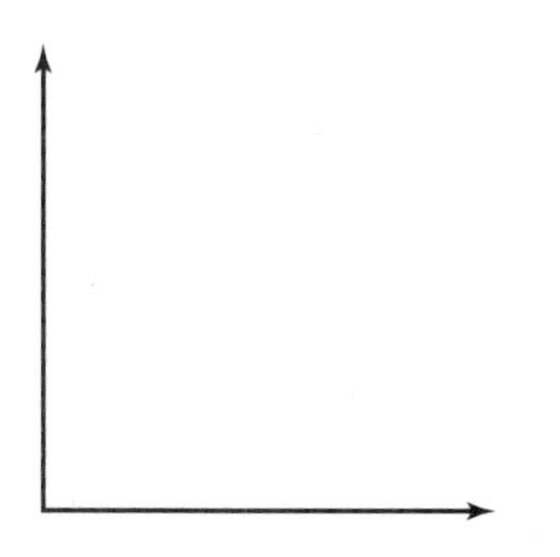

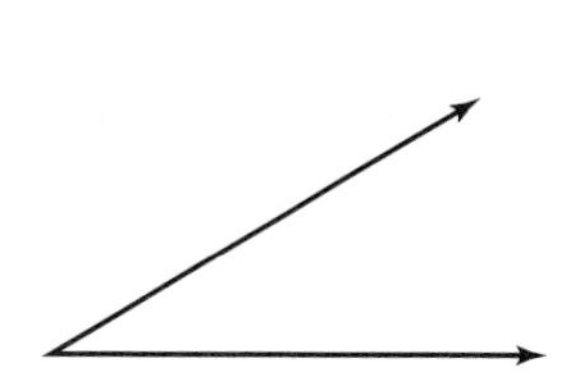

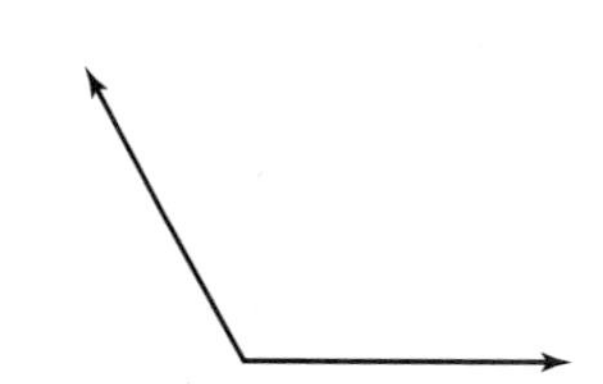

Right angles measure 90°.

Acute angles measure between 0° and 90°.

Obtuse angles measure between 90° and 180°.

EXAMPLE *Classify each angle.*

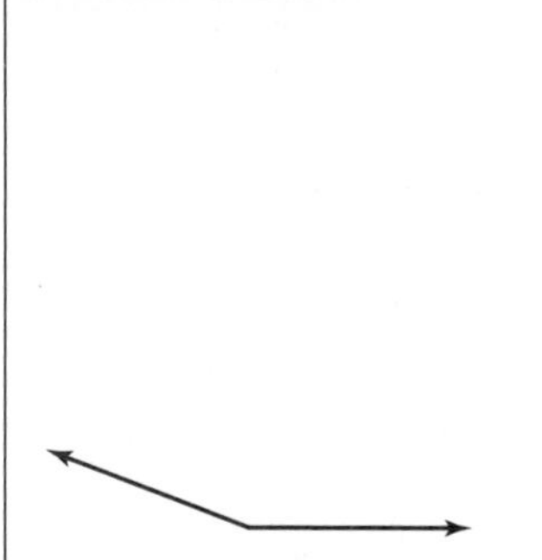

This angle is an obtuse angle.

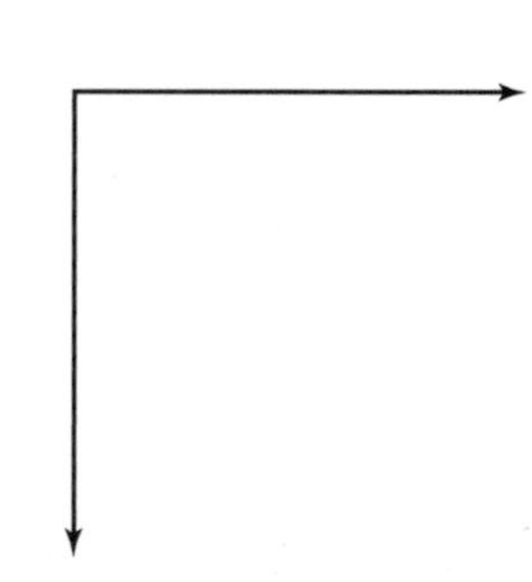

This angle is a right angle.

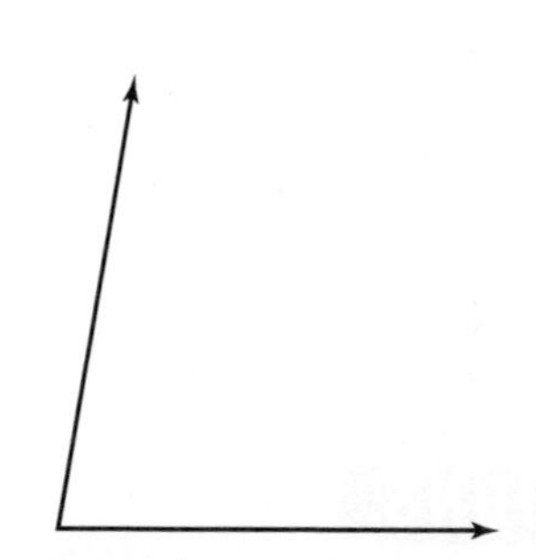

This angle is an acute angle.

EXERCISES *Classify each angle as right, acute, or obtuse.*

1. 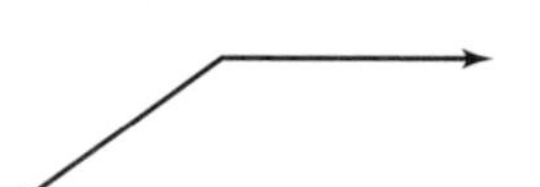

2. 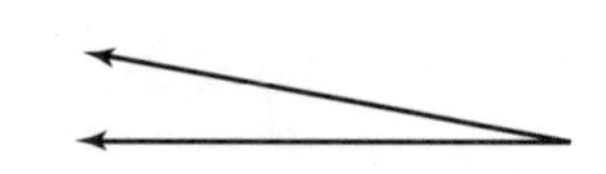

3. 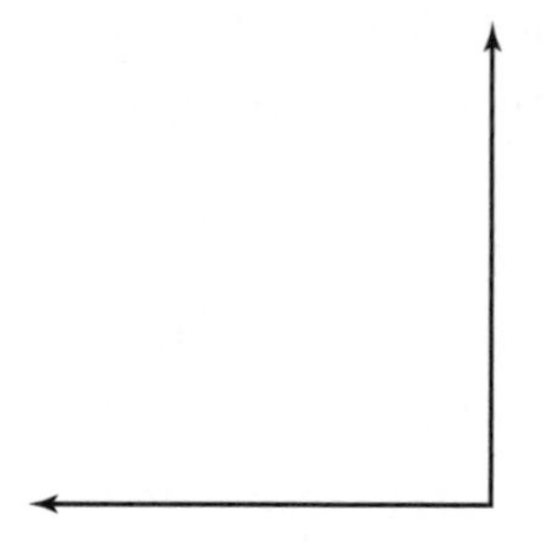

4. 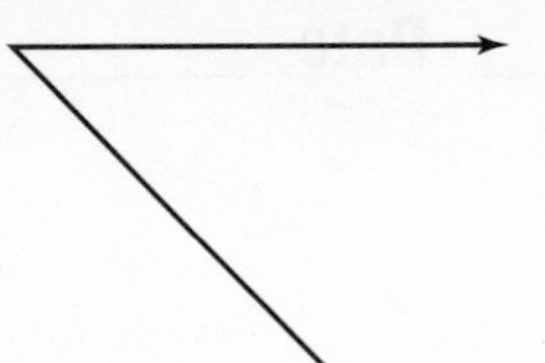

5. 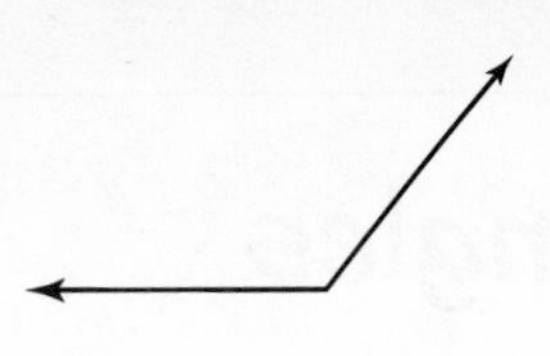

6. 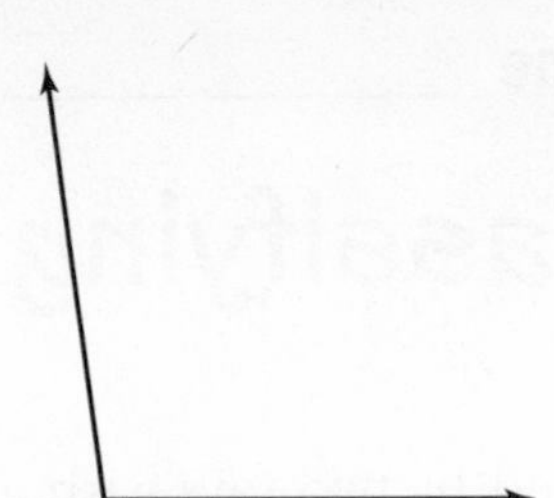

Classify the angles found in each polygon.

7. 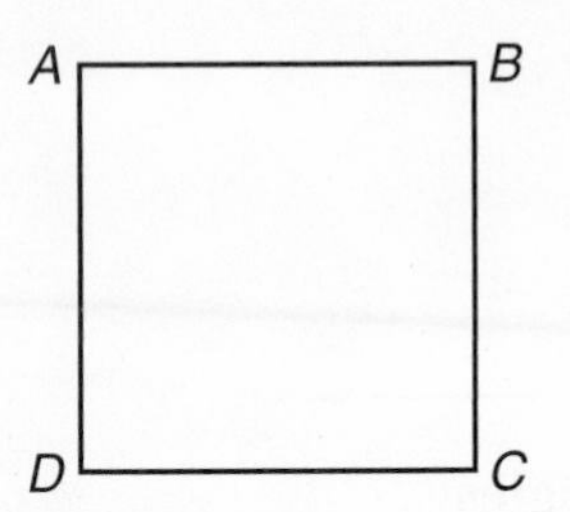

8. 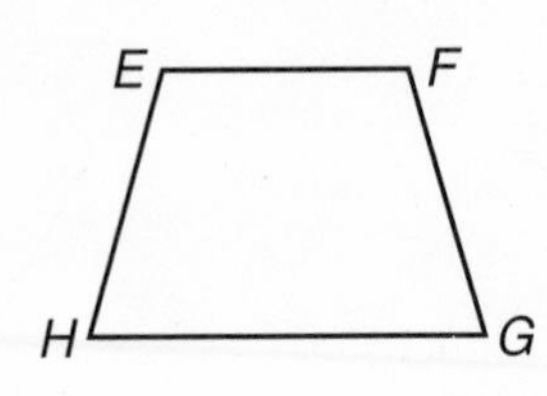

9. 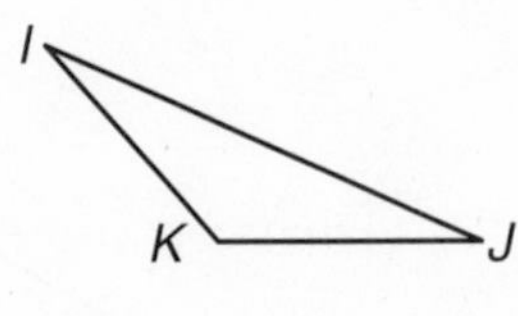

10. 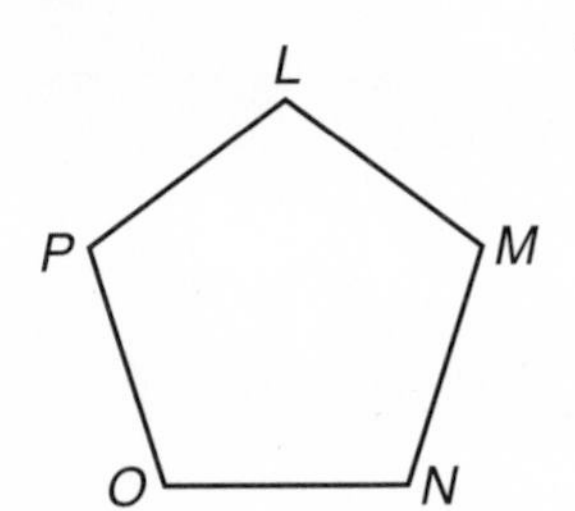

11. 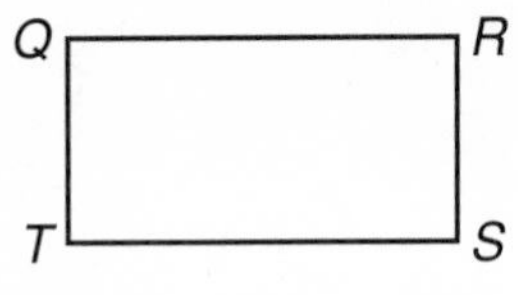

12. 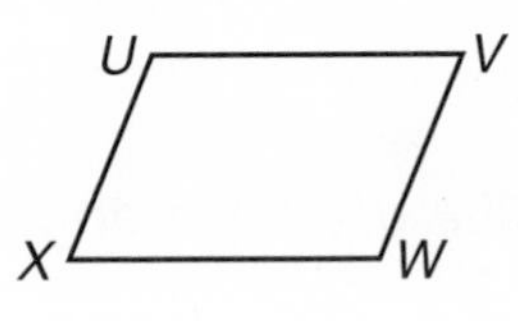

APPLICATIONS

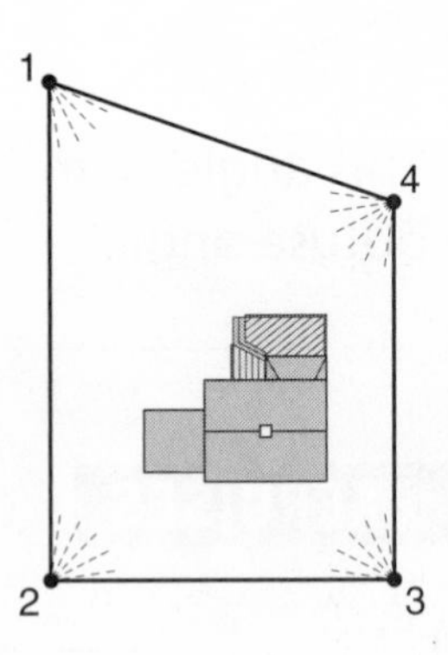

13. A diagram of Tara's lawn is shown at the right. Tara plans to place a sprinkler at each corner of the lawn. What type of angle should she set the spray for each sprinkler?

14. A diagram of a baseball field is shown at the right. What type of angle is formed from a ball thrown from first base to second base to third base?

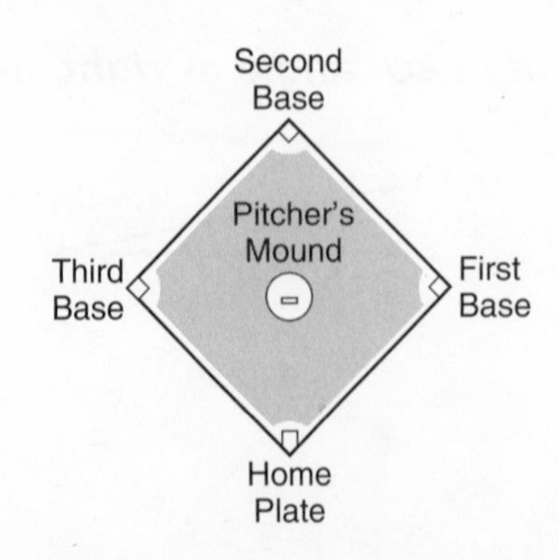

15. What type of angle is formed by a ball thrown from the pitcher to the catcher to the first baseman?

Name ______________________ Date __________

Polygons

Polygons are closed figures formed by line segments called sides. They are classified by their number of sides.

Polygon	Number of Sides
Triangle	3
Quadrilateral	4
Pentagon	5
Hexagon	6
Octagon	8

Regular polygons are polygons in which all sides are the same length and all the angles are the same size.

EXAMPLE ***Determine if each figure is a polygon. If the figure is a polygon, classify the polygon by the number of sides and as regular or not regular.***

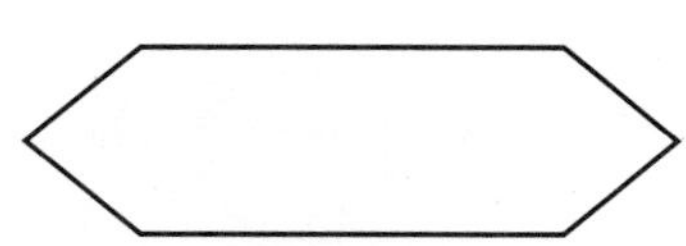

This figure is a polygon with six sides. It is a hexagon. It is *not* a regular polygon.

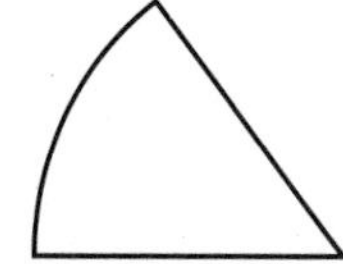

This figure is *not* a polygon since one of the sides is a curve.

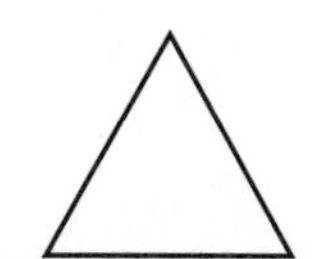

This figure is a polygon with three sides. It is a triangle. It is a regular polygon.

EXERCISES ***Tell whether each figure is a polygon. State* yes *or* no.**

1. 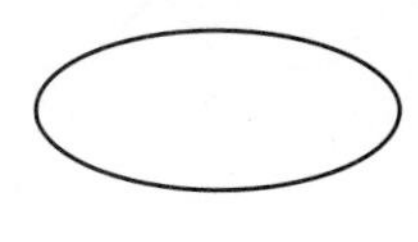

2. 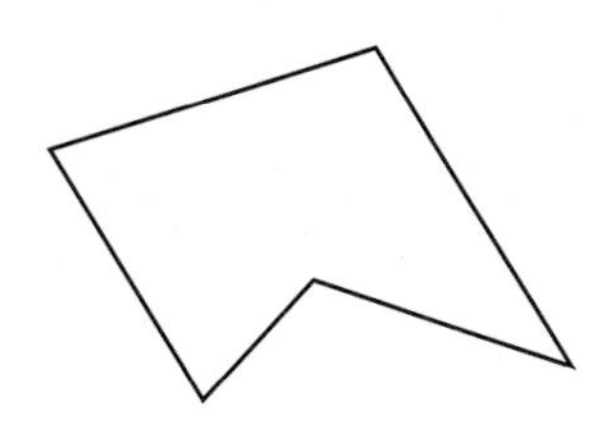

3. 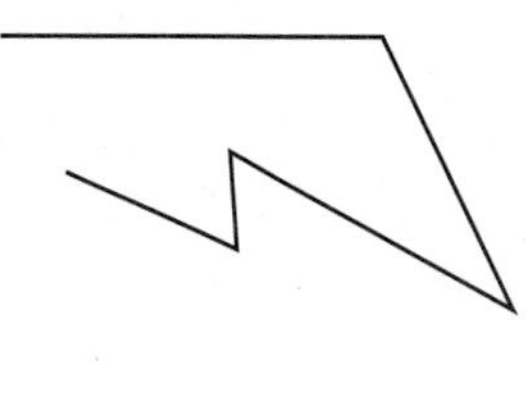

4. 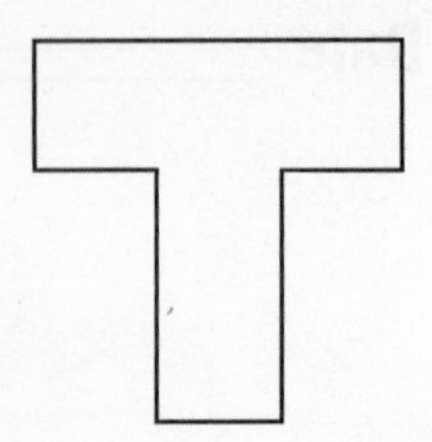

5. 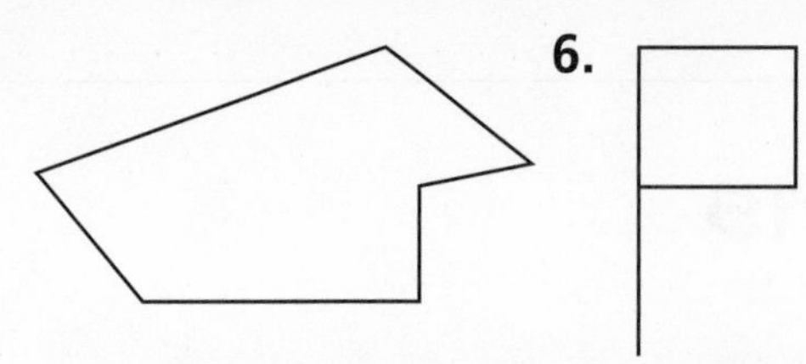

6.

Classify each polygon by the number of sides and as regular or not regular.

7. 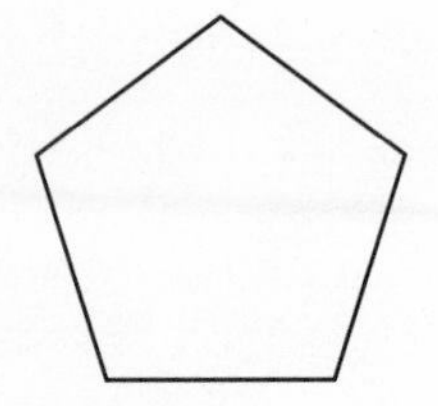

8. 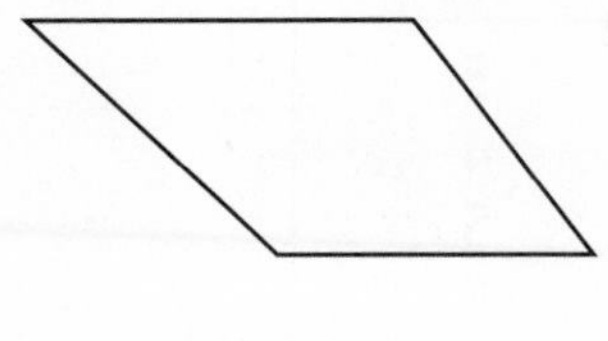

9. 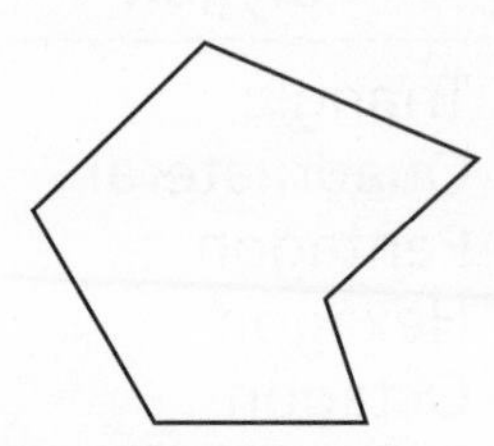

10. 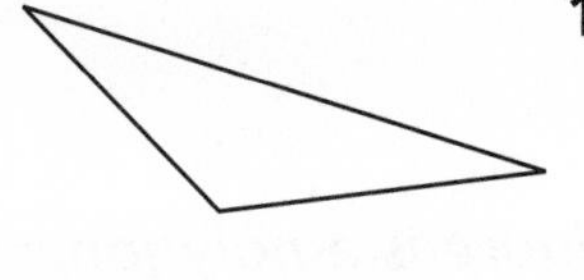

11. 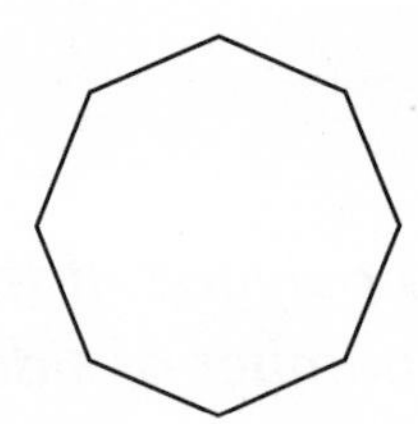

12. 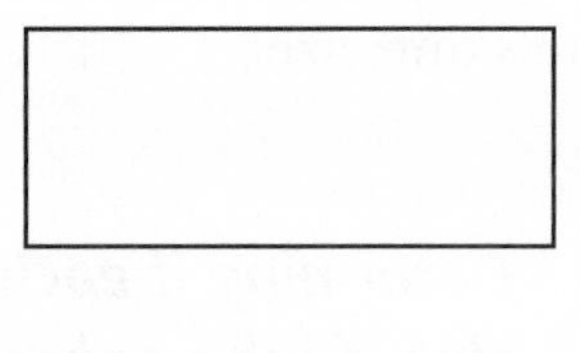

APPLICATIONS

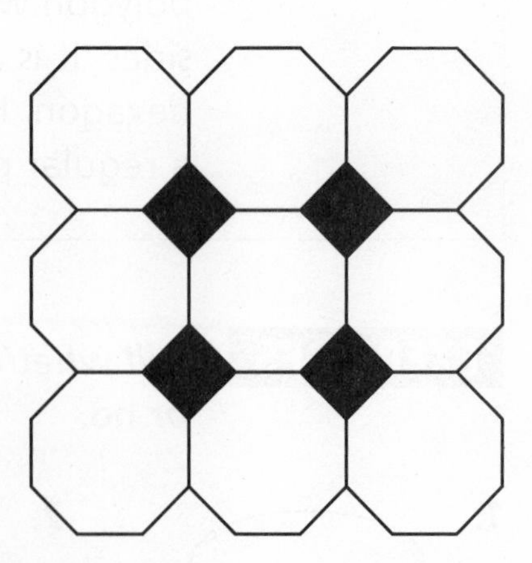

13. A tile pattern is shown at the right. Name the polygons in the pattern.

14. Find a picture of some interesting architecture. Name some examples of polygons in the picture.

15. Draw a picture or design by using various polygons.

Name ______________________ Date __________

Triangles and Quadrilaterals

Triangles may be classified by the measures of their angles or by the lengths of their sides.

Triangles			
Classification by Angles		**Classification by Sides**	
Acute	all angles acute	Scalene	all sides different lengths
Right	one right angle	Isosceles	two sides the same length
Obtuse	one obtuse angle	Equilateral	three sides the same length

Sides and angles are also used to classify quadrilaterals.

Quadrilaterals	
Trapezoid	only one pair of parallel sides
Parallelogram	both pairs of opposite sides parallel
Rectangle	parallelogram with four right angles
Rhombus	parallelogram with four sides the same length
Square	parallelogram with four right angles and four sides the same length

EXAMPLE ***Identify each polygon.***

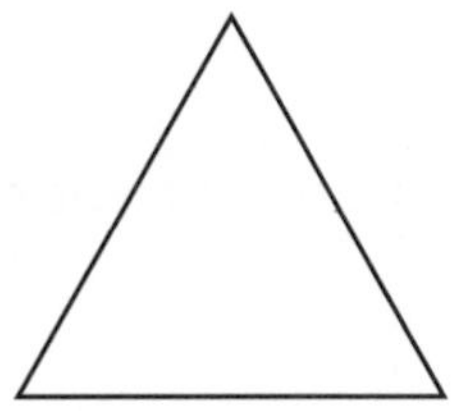

All of the angles are acute and all of the sides are the the same length. This triangle is acute and equilateral.

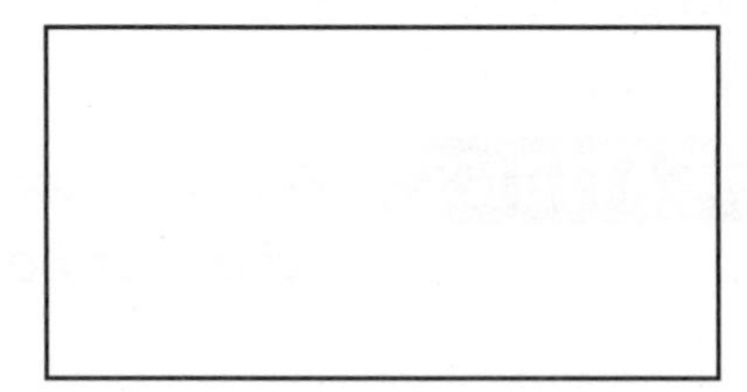

There are two pairs of parallel sides and four right angles. This quadrilateral is a rectangle.

EXERCISES *Classify each triangle by its sides and by its angles.*

1. 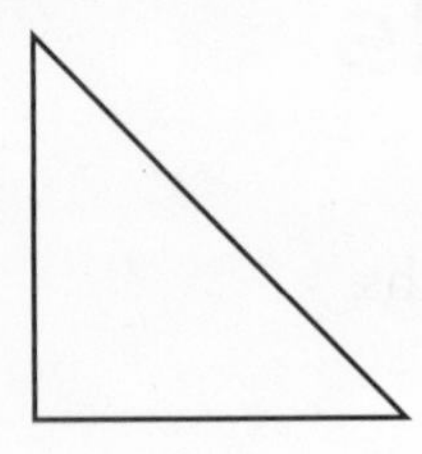

2. 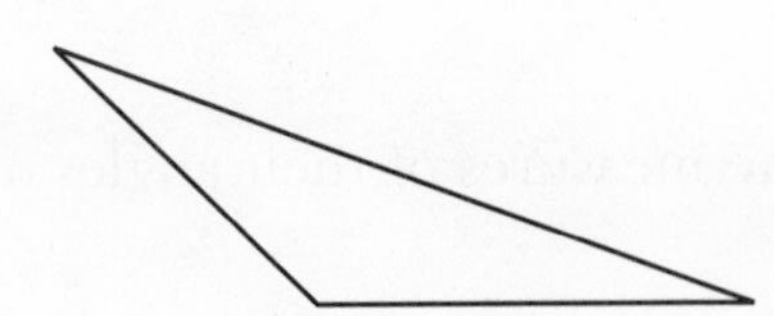

3. 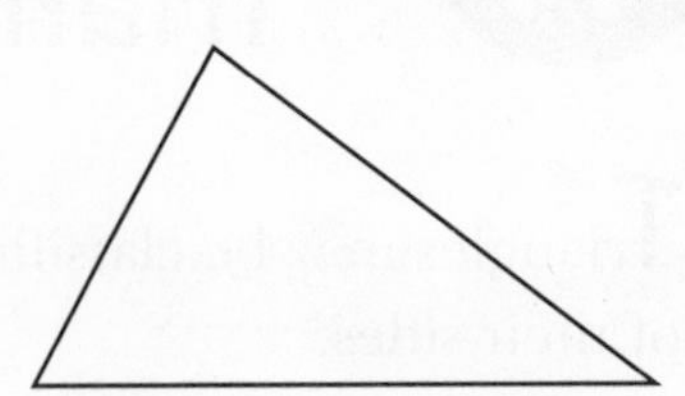

Name every quadrilateral that describes each figure. Then state which name best describes the figure.

4. 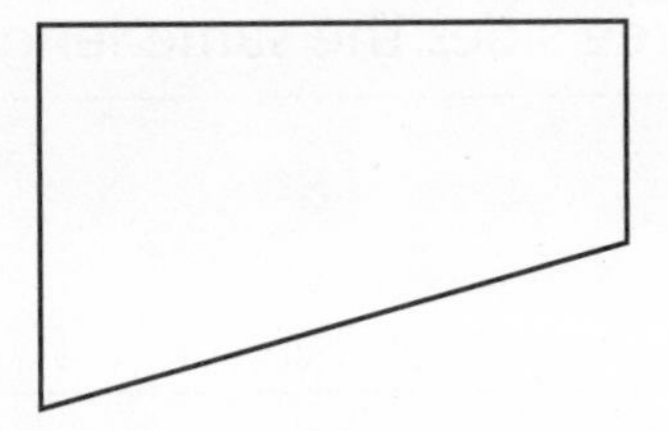

5. 

6. 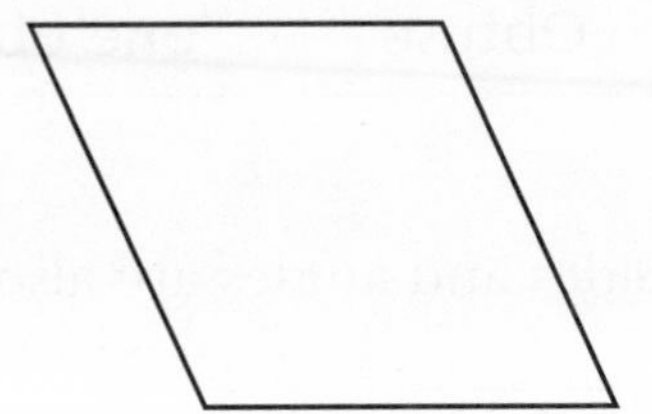

7. 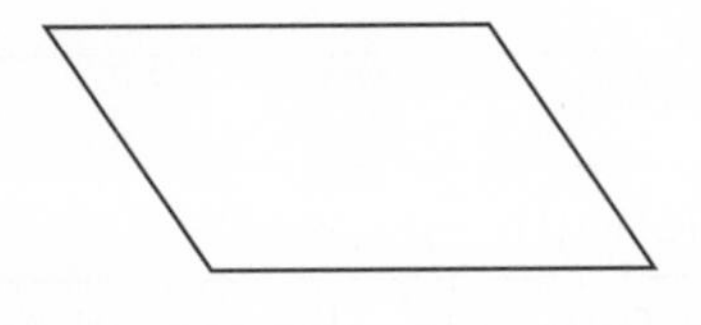

8. 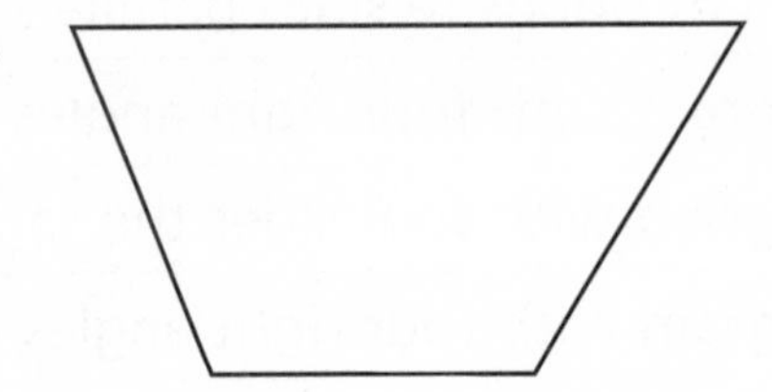

9. 

APPLICATIONS *Find two examples of each figure in your school or home.*

10. square
11. equilateral triangle
12. parallelogram
13. rectangle
14. right scalene triangle
15. trapezoid
16. acute isosceles triangle
17. rhombus
18. obtuse scalene triangle

Name ______________________________ Date __________

Line Symmetry

If a figure can be folded in half so that the two halves match exactly, the figure has a line of symmetry.

EXAMPLE *Draw all lines of symmetry for each figure.*

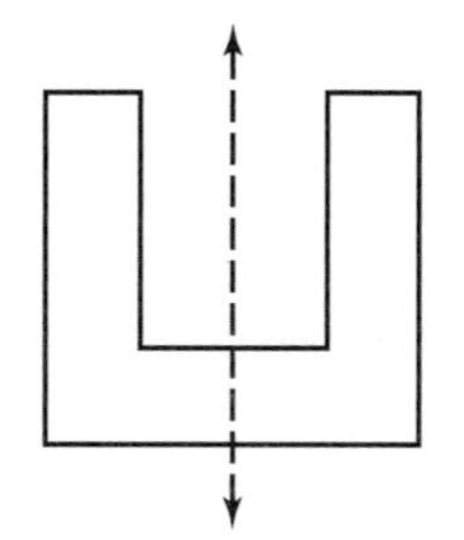
one line of symmetry

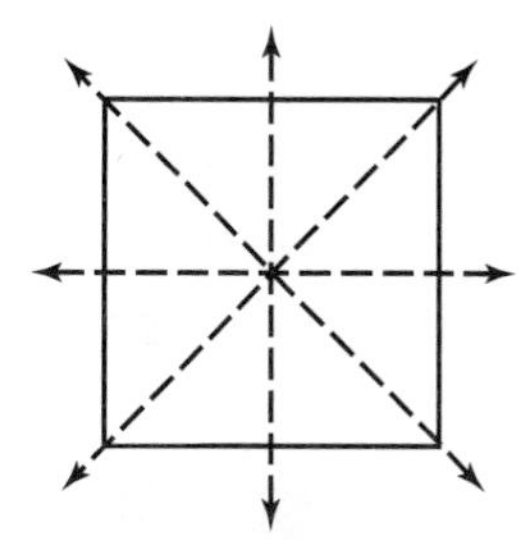
four lines of symmetry

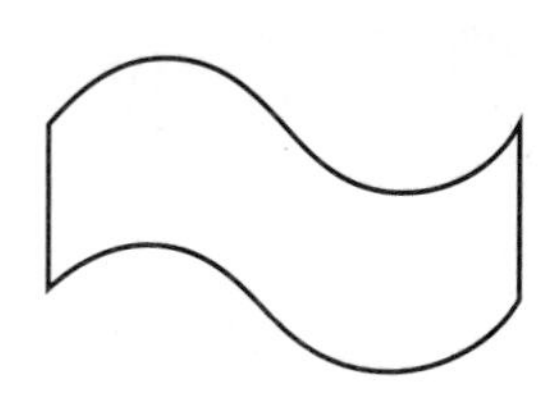
no lines of symmetry

EXERCISES *Draw all lines of symmetry for each figure.*

1.

2.

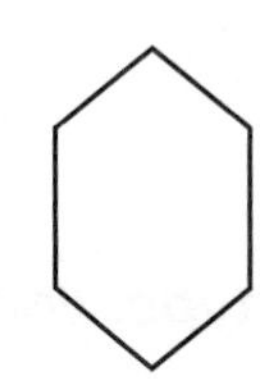

3.

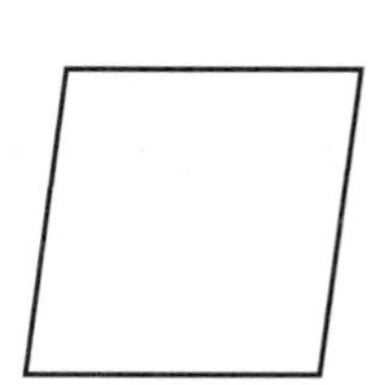

4.

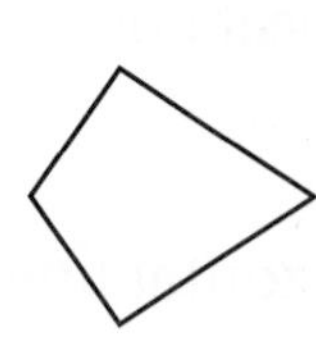

5.

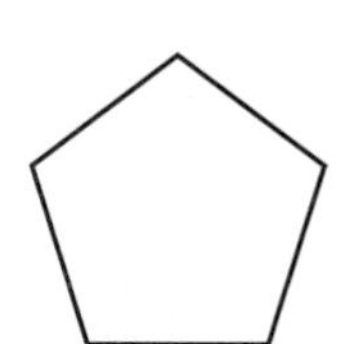

6.

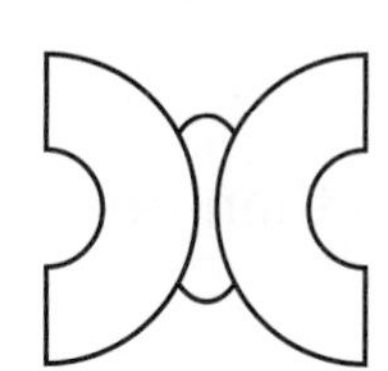

Complete each figure so that the dashed line is a line of symmetry.

7.

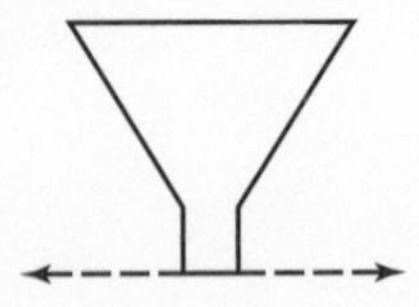

8.

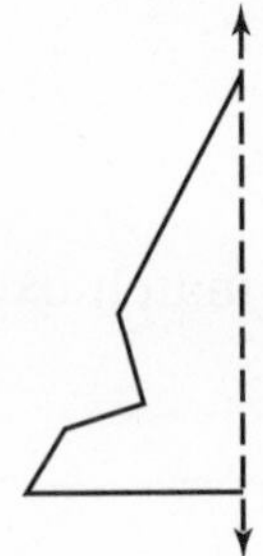

9.

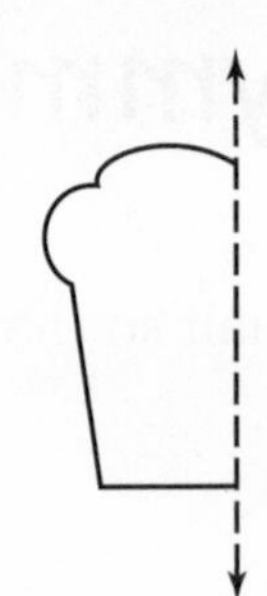

APPLICATIONS ***The following are designs from Navaho baskets. Determine the number of lines of symmetry for each of the designs.***

10.

11.

12.

Printers use many fonts or styles of type. For Exercises 13–16, consider block capital letters.

13. List the letters that have a vertical line of symmetry.

14. List the letters that have a horizontal line of symmetry.

15. List the letters that have no line of symmetry.

16. List the letters that have more than one line of symmetry.

Name ______________________ Date __________

Circumference of Circles

The formula for the **circumference** of a circle is $C = \pi d$ where C is the circumference and d is the diameter.

EXAMPLE ***The diameter of a Ferris wheel at the amusement park is 15 meters. How far does a seat on the Ferris wheel travel in one revolution?***

To find the distance traveled by a seat in one revolution, find the circumference of the Ferris wheel. Use the formula $C = \pi d$. Substitute 3.14 for π and 15 for d.

$$C = \pi d$$
$$C \approx 3.14 \cdot 15$$
$$C \approx 47.1$$

The circumference of the Ferris wheel is about 47.1 meters. Thus, a seat travels about 47.1 meters in one revolution.

EXERCISES ***In Exercises 1–12, find the circumference of each circle shown or described. Use 3.14 for π.***

1.
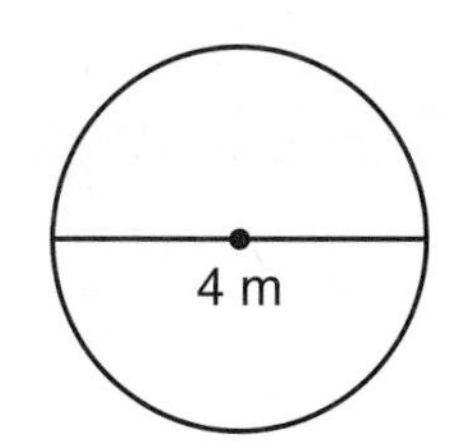

2.
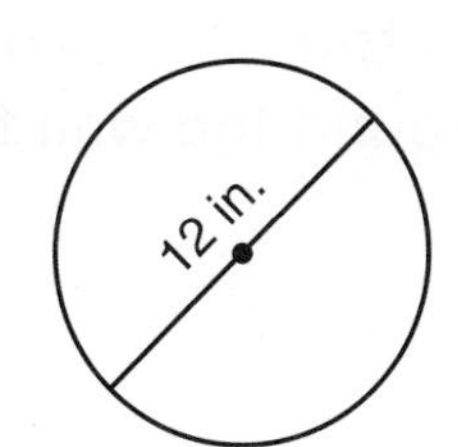

3.
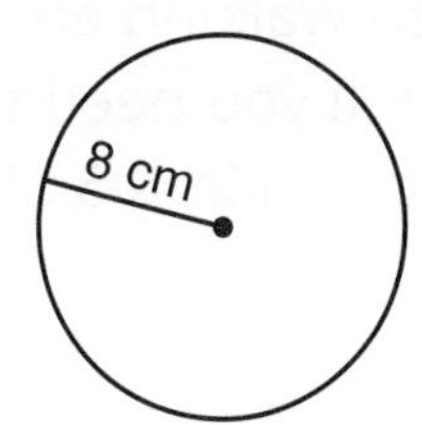

4.
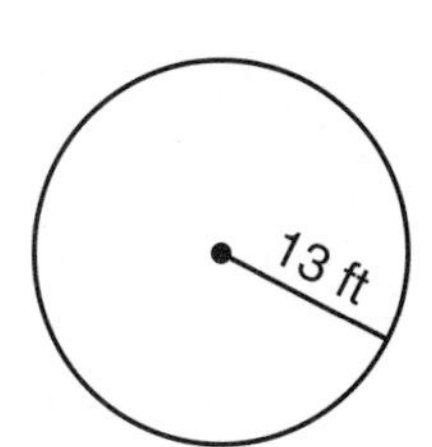

5.
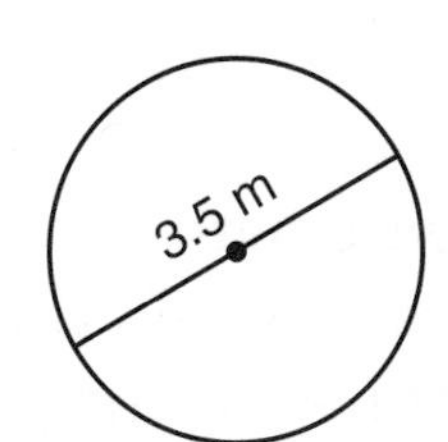

6.
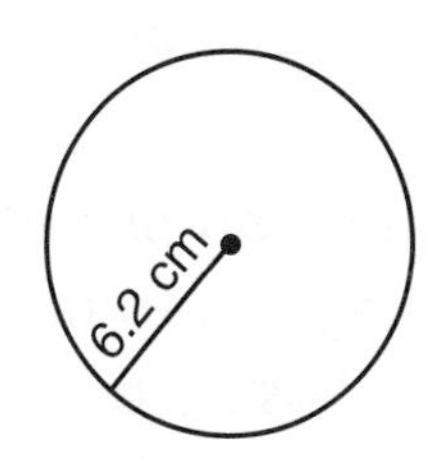

7. The diameter is 16.4 km.

8. The radius is 0.5 m.

9. The radius is 17 ft.

10. The diameter is 4.7 in.

11. The radius is 1.3 cm.

12. The diameter is 10 in.

APPLICATIONS ***The Castle Garden is a national monument on Manhattan Island in New York City. It was originally built by the Dutch in the seventeenth century to be used as a fort. It has a diameter of about 236 feet. Use this information to answer Exercises 13–15.***

13. Suppose you are standing in the center of Castle Garden and you walk toward a wall. How far will you walk?

14. You decide to walk completely around the outside wall. How far will you walk, to the nearest foot?

15. You decide to keep walking around the outside wall. About how many times will you need to walk around the wall to walk 1 mile? (Hint: 1 mile = 5,280 feet)

16. The distance around Earth at the equator is about 25,000 miles. What is the approximate diameter of Earth at the equator, to the nearest mile?

17. Ted's town is planning on putting in a bicycle path at the local park. They want the path to be 400 meters long and circular. What should the radius be of the circle formed by the path, to the nearest meter?

Name ____________________ Date __________

Perimeter of Rectangles, Squares, and Parallelograms

The **perimeter** of a figure is the distance around the figure.

EXAMPLE ***Determine how much fence will enclose a rectangular yard with a length of 50 feet and a width of 63 feet.***

To find the perimeter of a figure, add the measures of its sides.

$$P = 50 + 50 + 63 + 63 \text{ or } 226$$

The perimeter or amount of fence needed is 226 feet.

EXERCISES ***Find the perimeter of each figure shown or described below.***

1.

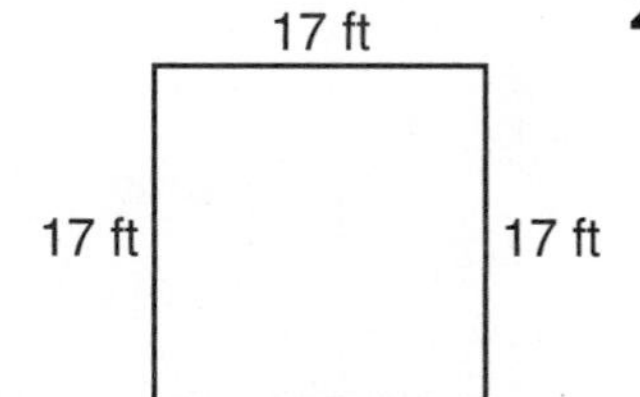

2.

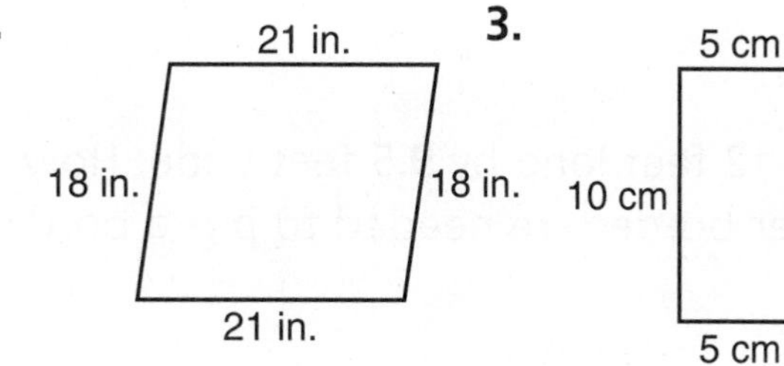

3.

5 cm

10 cm 10 cm

5 cm

4. rectangle:

$\ell = 3.5$ m
$w = 4$ m

5. square:

$s = 2.5$ in.

6. rectangle:

$\ell = 12.75$ ft
$w = 8.5$ ft

Measure each figure to find the perimeter. Measure to the nearest fourth inch.

7.

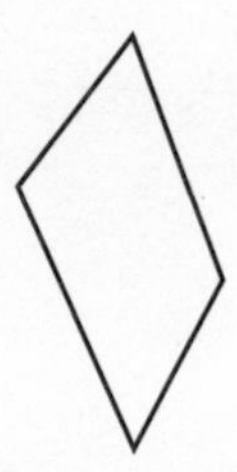

8.

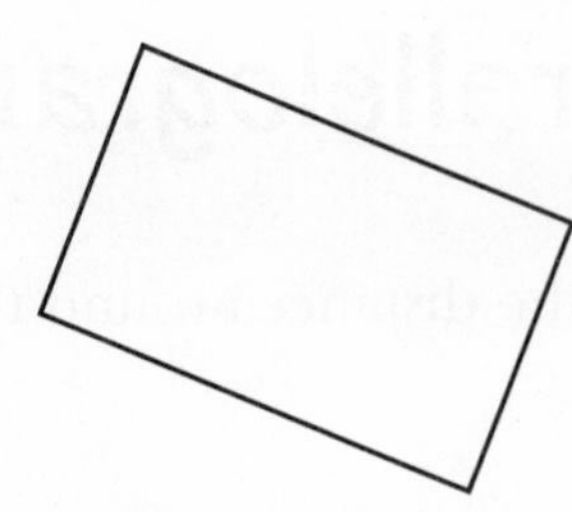

9.

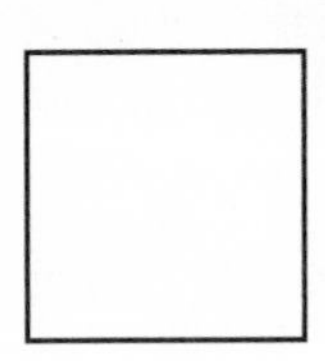

10.

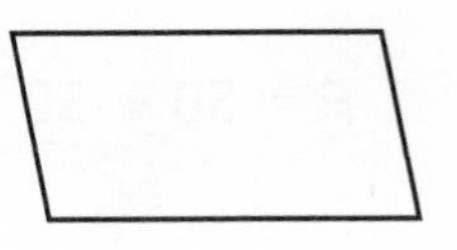

APPLICATIONS

11. A rectangular room is 12 feet long by 9.5 feet wide. How many feet of wallpaper border are needed to put a border around the room?

12. Mr. Nichols wants to enclose a rectangular garden with wire fencing. The garden is against his garage so he needs to fence only three sides. How much fence does he need if the perimeter of the whole garden is 86 feet and the side of the garage is 25 feet? Draw and label a diagram of the garden.

13. Draw and label a parallelogram that has a perimeter of 30 inches.

Name ______________________ Date __________

Area of Circles

The area of a circle is given by the formula $A = \pi r^2$ where A is the area and r is the radius.

EXAMPLE ***Gwen is making a circular rug with a radius of 3 feet. What will the area of the rug be that she is making?***

$A = \pi r^2$
$A \approx 3.14 \cdot 3^2$ *Substitute 3.14 for* π *and 3 for r.*
$A \approx 3.14 \cdot 9$
$A \approx 28.26$

The area of the rug is about 28.26 square feet.

EXERCISES ***In Exercises 1–12, find the area of each circle shown or described. Use 3.14 for*** π.

1.

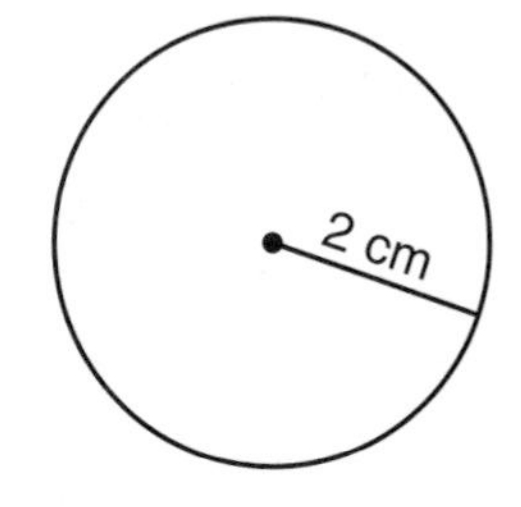

2.

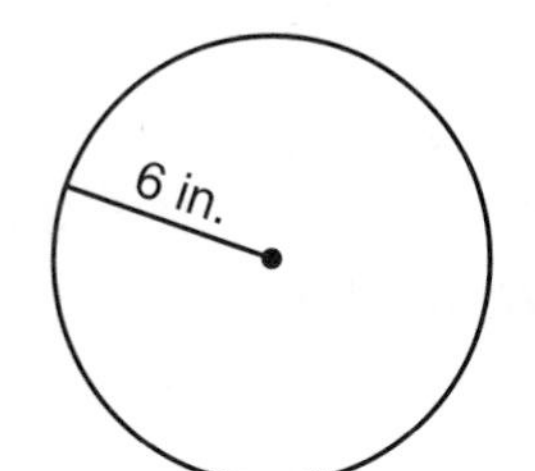

3.

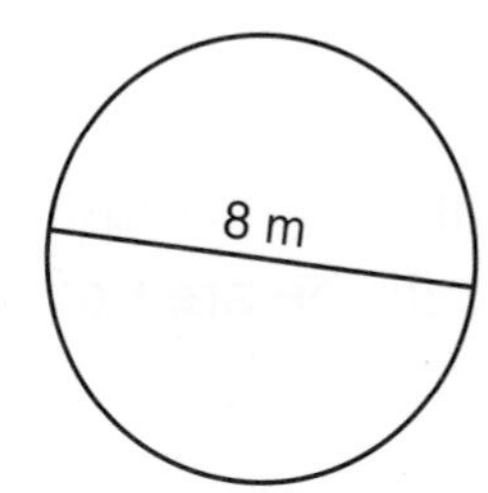

4.

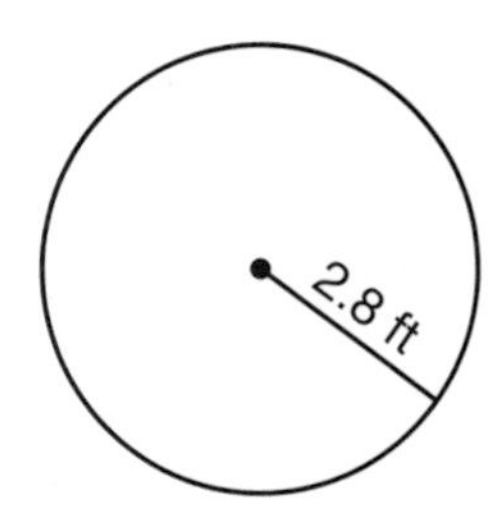

5.

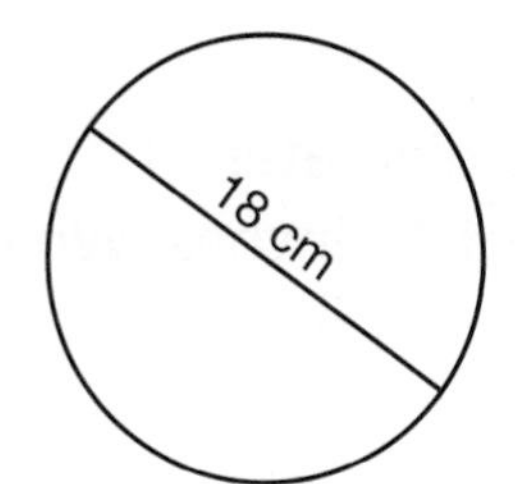

6.

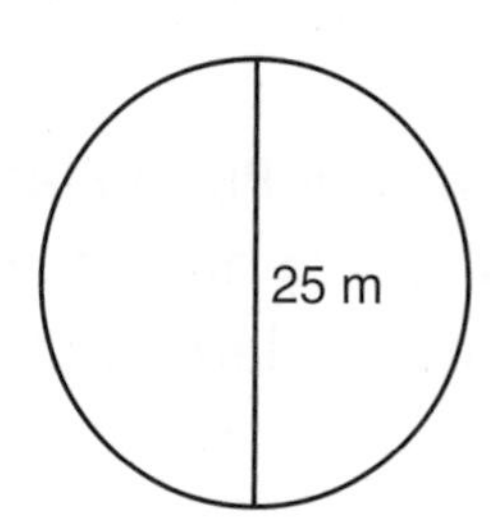

7. The radius is 5 m.

8. The radius is 10 ft.

9. The radius is 3.6 cm.

10. The diameter is 5 in.

11. The radius is 8.4 km.

12. The diameter is 4.6 yd.

APPLICATIONS ***Joani is preparing a circular garden with a diameter of 24 feet. Use this information to answer Exercises 13–16.***

13. What is the area of the garden?

14. She wants to cover the entire area with peat moss. If each bag of peat moss covers 160 square feet, how many bags of peat moss will she need?

15. Next year, she plans on increasing the diameter of the garden by 2 feet. What will the area of the new garden be?

16. How many bags of peat moss will she need to cover the new garden?

17. A large pizza from The Pizza Place has a diameter of 16 inches. A small pizza has a diameter of 10 inches. Which has the greater area, 1 large pizza or 2 small pizzas?

Name ______________________ Date __________

Area of Rectangles, Squares, and Parallelograms

Area is the number of square units needed to cover a surface.

Figure	Rectangle	Square	Parallelogram
Area Formula	$A = (\ell)(w)$ (ℓ = length) (w = width)	$A = s^2$ (s = side)	$A = (b)(h)$ (b = base) (h = height)
Example	18 in. 20 in. $A = (\ell)(w)$ $A = (20)(18)$ $A = 360$ The area is 360 square inches.	s = 4.5 ft $A = s^2$ $A = 4.5^2$ $A = 20.25$ The area is 20.25 square feet.	2 m 12 m $A = (b)(h)$ $A = (12)(2)$ $A = 24$ The area is 24 square meters.

EXERCISES ***In Exercises 1–9, find the area of each figure shown or described.***

1.

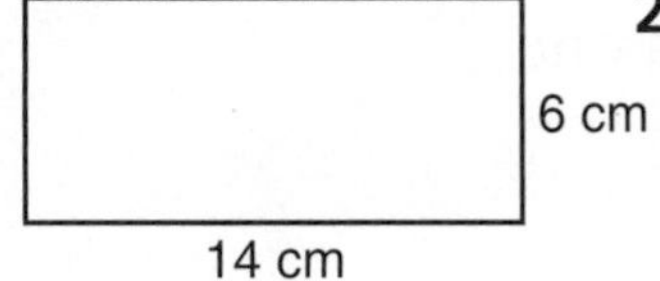

2.

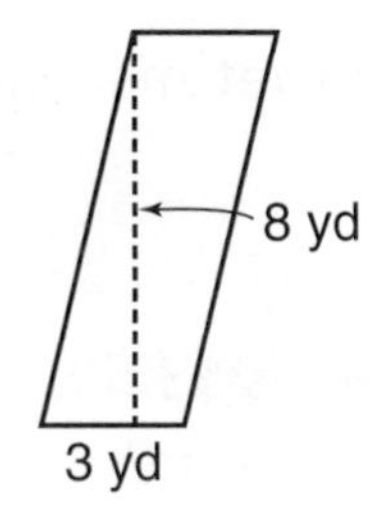

3.

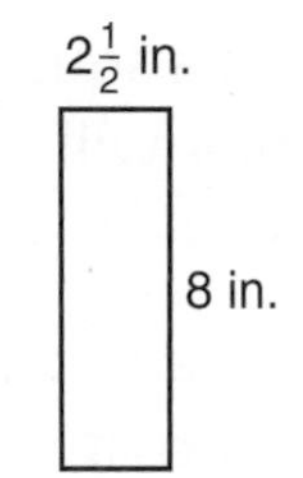

4. parallelogram
$b = 15$ ft
$h = 21$ ft

5. rectangle
$\ell = 7.5$ cm
$w = 12$ cm

6. parallelogram
$b = 4.7$ m
$h = 2.2$ m

7.

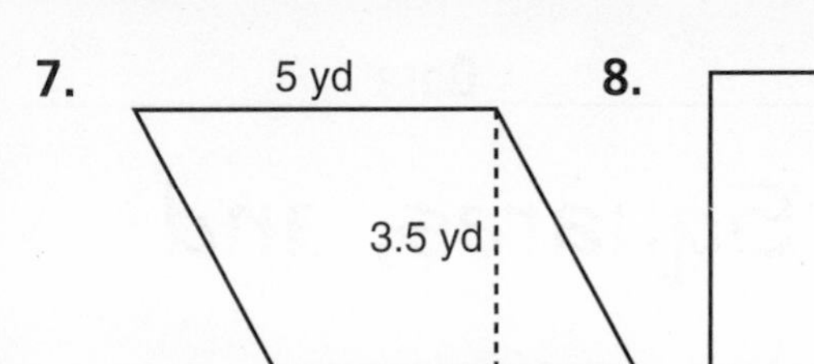

8.

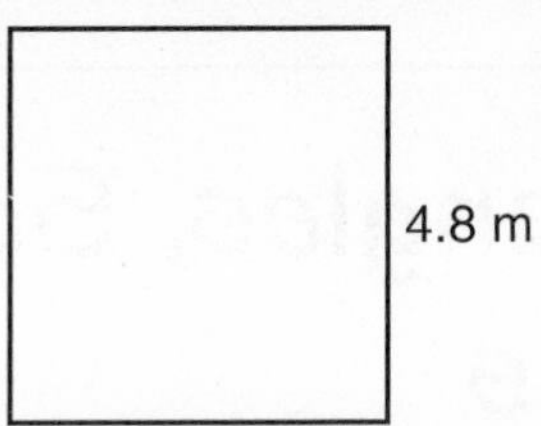

9.

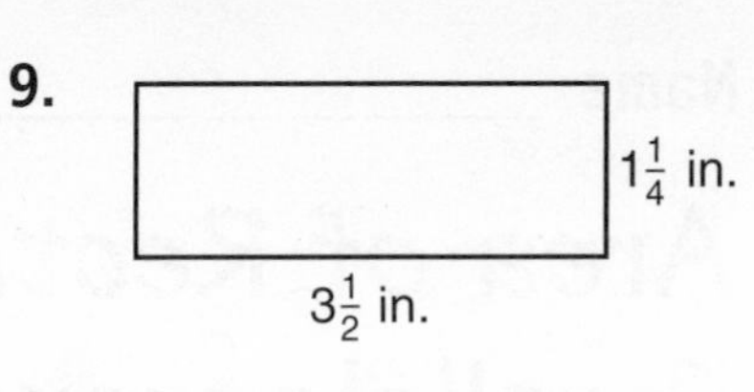

10. Find the area of a regulation-size volleyball court with a length of 59 feet and a width of 29.5 feet.

11. Find the length of a rectangle with an area of 84 square inches and a width of 7 inches.

12. Find the height of a parallelogram with a base of 12 yards and an area of 39 square yards.

APPLICATION

13. The figure at the right is the floor plan of a family room. The plan is drawn on grid paper, and each square of the grid represents one square foot. The floor is going to be covered completely with tiles.

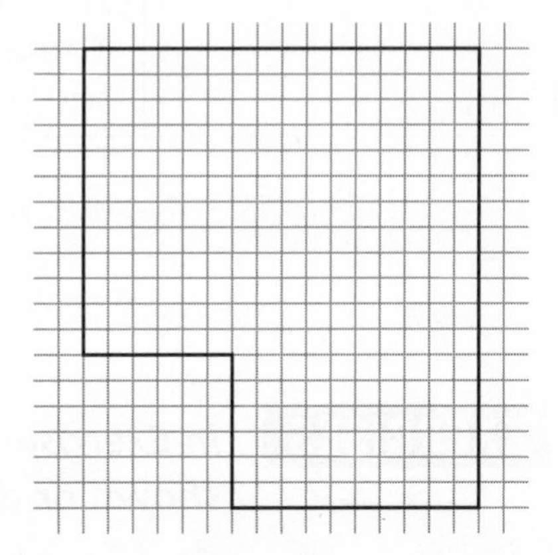

a. What is the area of the floor?

b. Suppose each tile is a square with a side that measures one foot. How many tiles will be needed?

c. Suppose the cost of a 1 foot by 1 foot tile is $3.50. How much would it cost to tile the entire floor?

d. Suppose the 1 foot by 1 foot tile had a cost of two tiles for $6.99. How much would it cost to tile the entire floor?

SKILL 50

Name ______________________ Date __________

Area of Triangles

The area of a triangle is equal to one half the product of its base (b) and height (h).

$$A = \frac{1}{2}bh$$

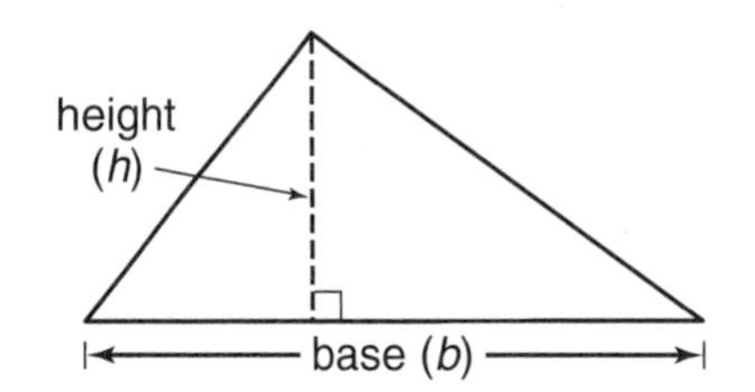

EXAMPLE ***Find the area of the triangle shown at the right.***

$A = \frac{1}{2}bh$

$A = \frac{1}{2} \times 15 \times 26$

$A = 195$

The area of the triangle is 195 square feet.

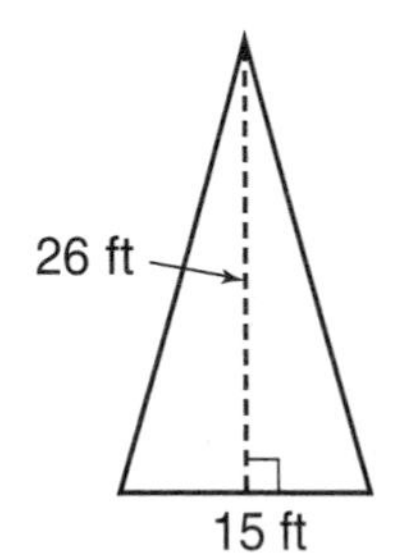

EXERCISES ***Find the area of each triangle.***

1. base, 12 inches
height, 7 inches

2. base, 20 centimeters
height, 12 centimeters

3. base, 8 feet
height, 24 feet

4. base, 17 meters
height, 6 meters

5. base, 6 kilometers
height, 13 kilometers

6. base, 10 yards
height, 20 yards

7.

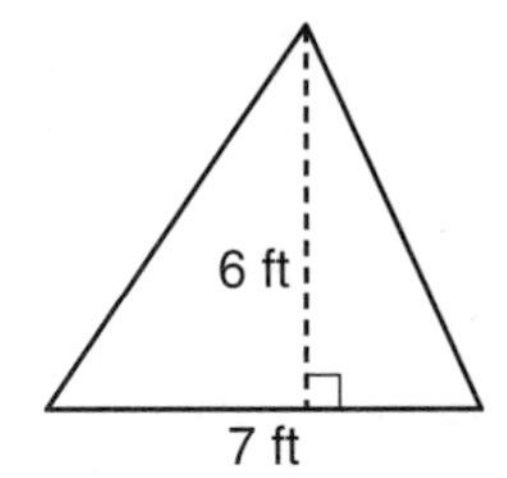

8.

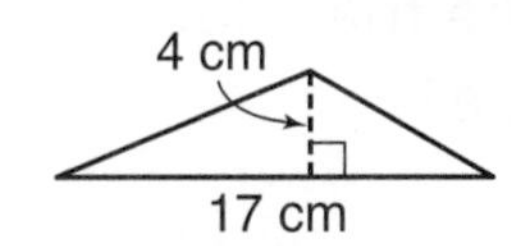

9.

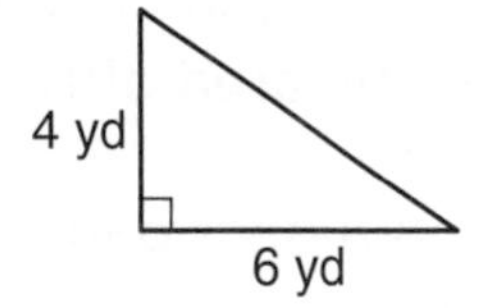

10.

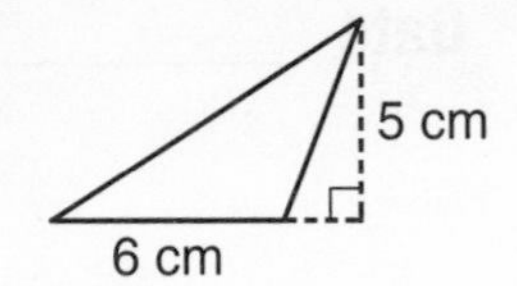

11.

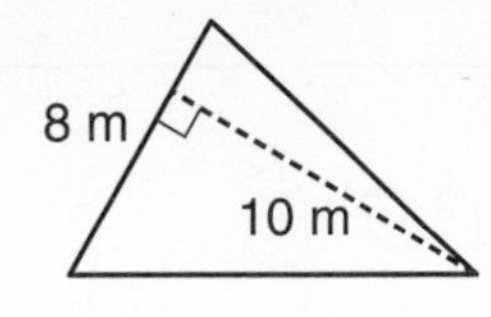

12.

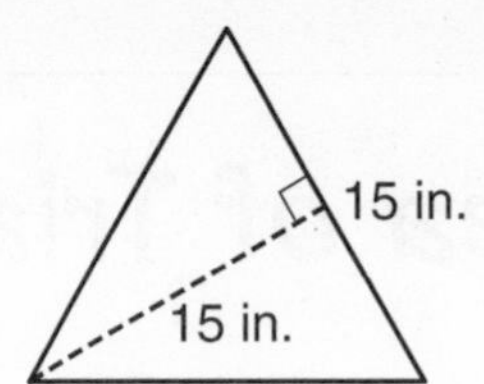

13.

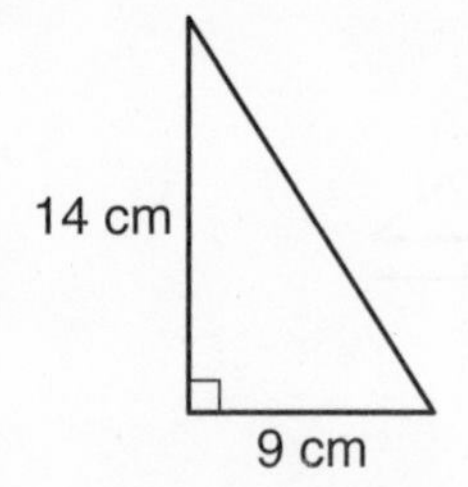

14.

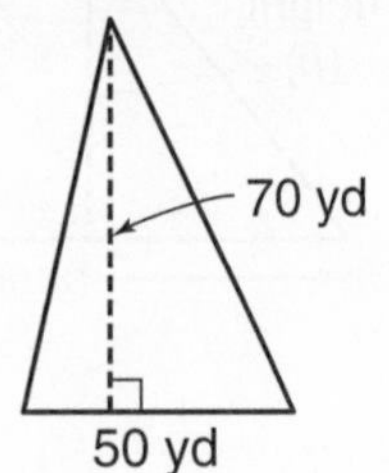

15.

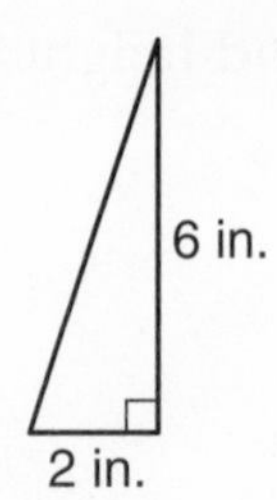

APPLICATIONS

16. Tom Wise has an A-frame cabin. What is the area of the front of the home?

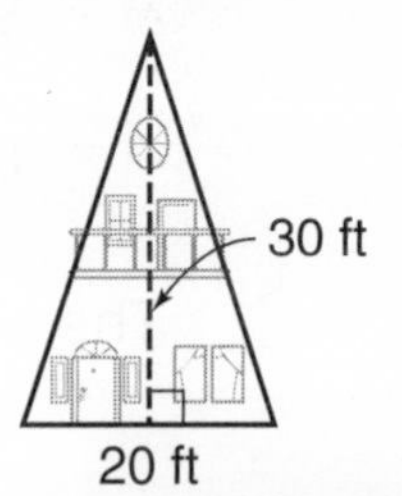

17. Tammy's bedroom is 12 feet by 12 feet. She plans to divide the room into two halves along the diagonal. She plans to carpet one half of the room with black carpet and the other half with white carpet. How many square feet of black carpet will she need?

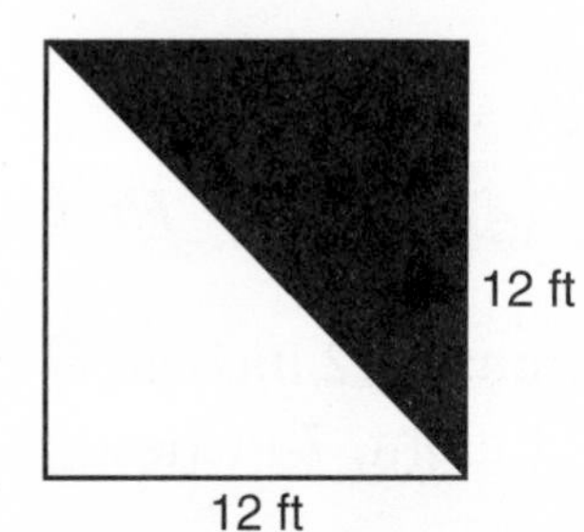

18. Use a geoboard or dot paper to make the triangle at the right. What is the area of the triangle?

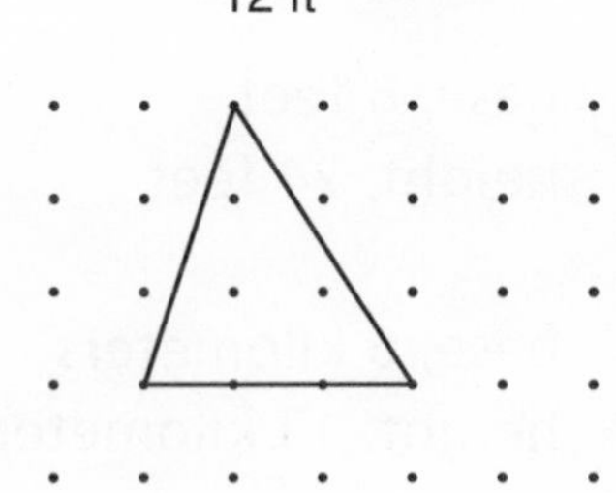

19. Use a geoboard or dot paper to make the triangle at the right. What is the area of the triangle?

20. Make a triangle on a geoboard or dot paper. Find the area of the triangle.

21. Make a triangle on a geoboard or dot paper that has an area of 8 square units.

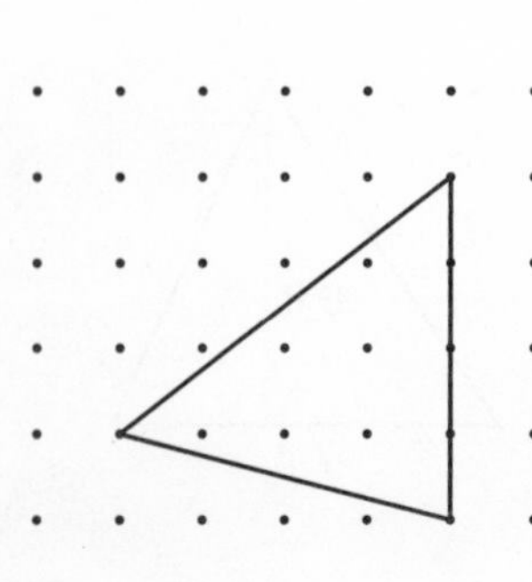

Name ______________________ Date __________

Area of Trapezoids

A trapezoid is a quadrilateral with exactly one pair of parallel sides. The area of a trapezoid is equal to the product of half the height and the sum of the bases.

$$A = \frac{1}{2}h(a + b)$$

EXAMPLE ***Find the area of the trapezoid shown at the right.***

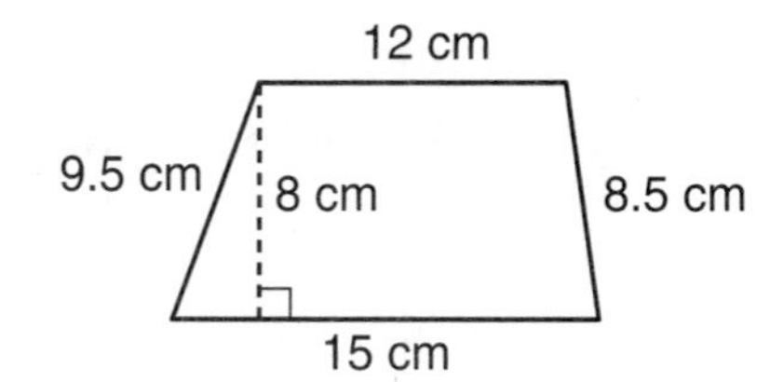

$A = \frac{1}{2}h(a + b)$

$A = \frac{1}{2}(8)(15 + 12)$

$A = \frac{1}{2}(8)(27)$

$A = 108$

The area of the trapezoid is 108 square centimeters.

EXERCISES ***In Exercises 1–6, find the area of each figure shown or described.***

1.

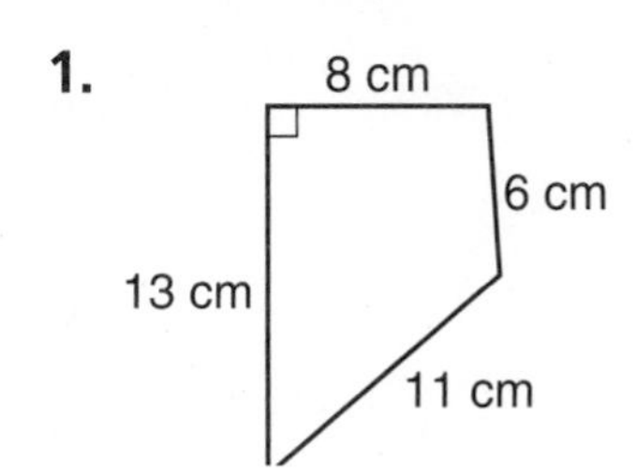

2.

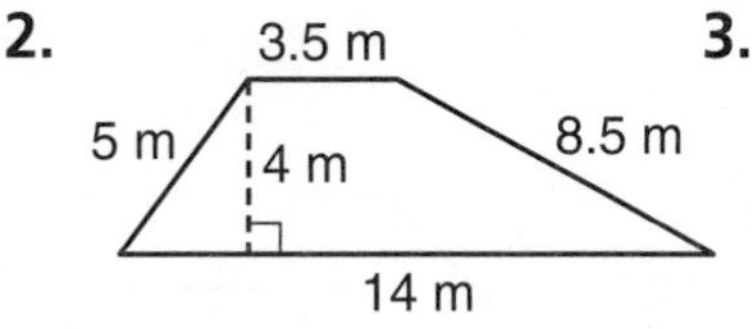

3.

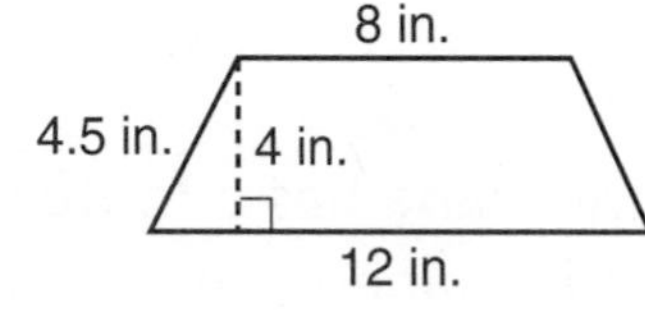

4. bases: 6 m, 9 m
height: 4 m

5. bases: 10 ft, 15 ft
height: 20 ft

6. bases: 7.6 cm, 10 cm
height: 8 cm

7. Find the area of the figure at the right.

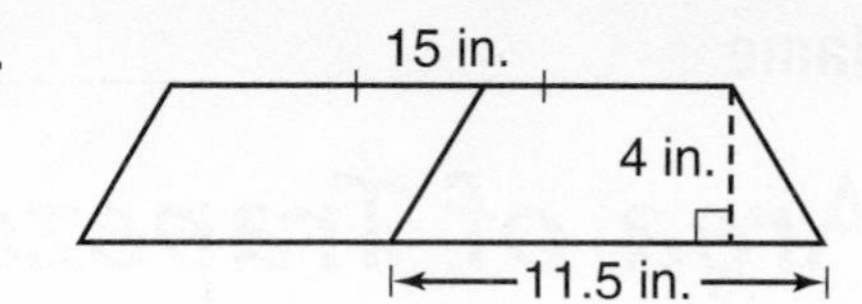

APPLICATIONS

8. Jose received a new entertainment center for his bedroom. It holds his TV, VCR, stereo, CD player, and tapes. It sits on a pedestal base that is the shape of a trapezoid. The two bases of the trapezoid are 36 inches and 28 inches long, and the height is 12 inches. What is the area of the front of the base?

9. Use the figure at the right to determine the area of the three trapezoidal-shaped spaces between the steps of the 4-foot ladder. The bottom base lengths are given for each space. The top base lengths are 2 inches shorter than the bottom base lengths. Each step between the spaces is 4 inches high and the spaces (including the bottom space) are all the same height.

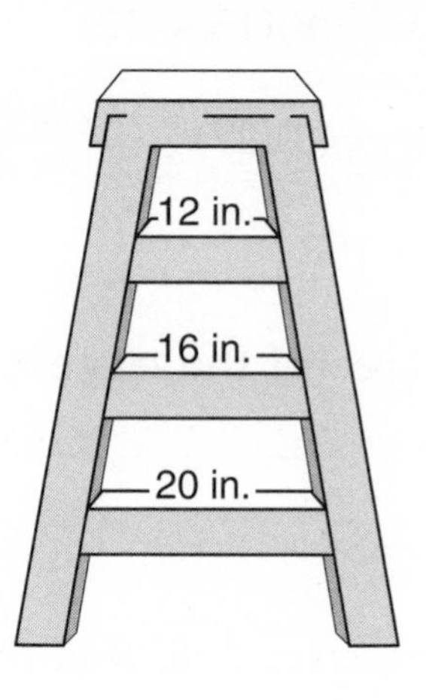

10. Columbus, Montana has a football stadium that is shaped like the figure at the right. The center part is the field, and the outside part is the seats. The longest field length is 140 yards, and the longest field width is 100 yards. Find the area of the field.

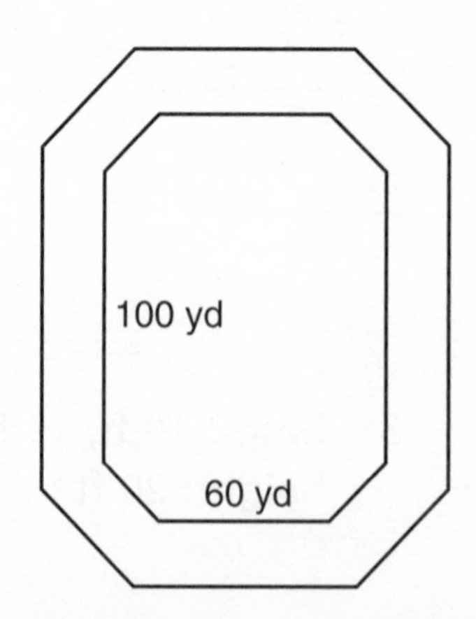

SKILL 52

Name ______________________ Date ________

Three-Dimensional Figures

Drawings of some common three-dimensional figures are shown below.

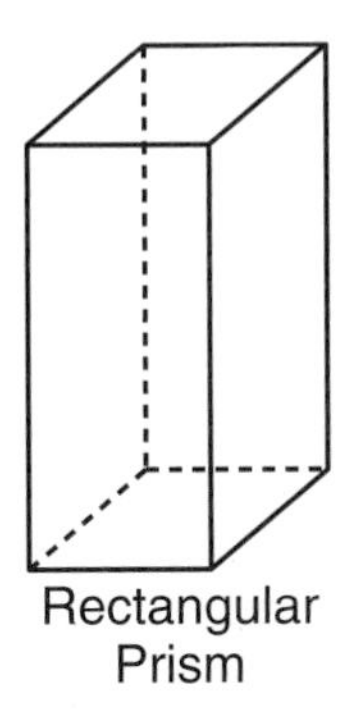
Rectangular Prism

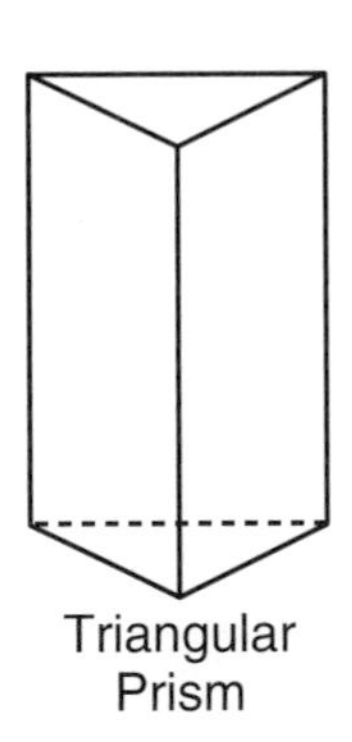
Triangular Prism

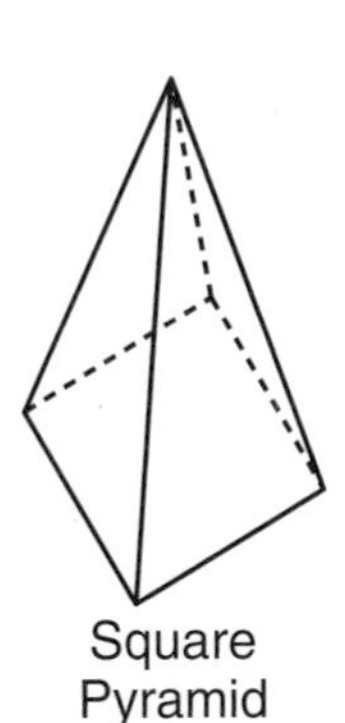
Square Pyramid

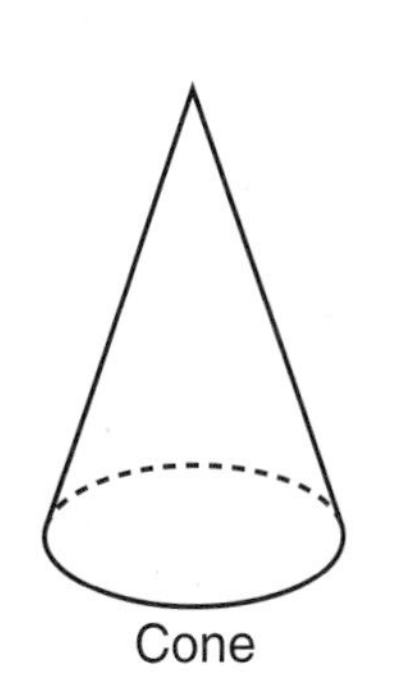
Cone

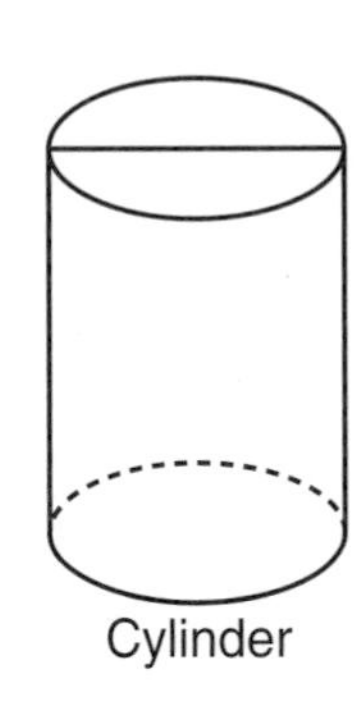
Cylinder

The flat surfaces of a prism or a pyramid are called **faces**. The faces intersect to form **edges**. The edges intersect to form **vertices**.

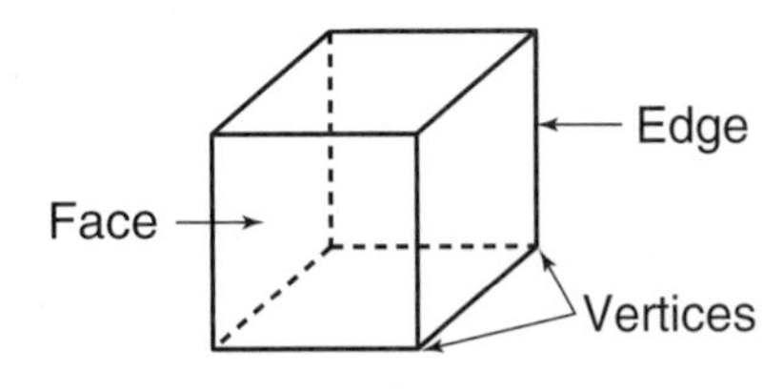

EXAMPLE ***Draw a hexagonal prism.***

Step 1 Lightly draw a hexagon for a base.

Step 2 Lightly draw lines straight up from each vertex of the hexagon. Make each line the same length.

Step 3 Draw the top hexagon.

Step 4 Go over the lines you have drawn. Make the edges you can see solid lines and the edges you cannot see dashed lines.

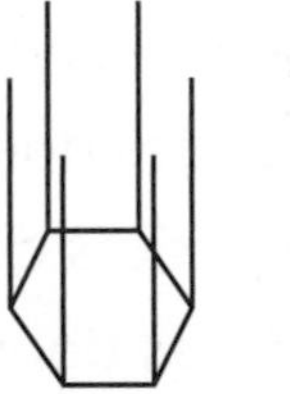

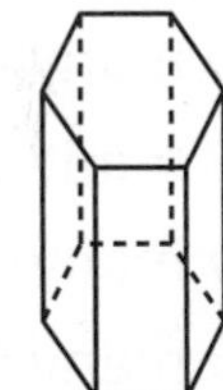

EXERCISES ***Draw each of the following.***

1. a triangular prism

2. a hexagonal pyramid

3. a rectangular prism

4. a triangular pyramid

5. a cylinder

6. a cone

Use your drawings to answer each question.

7. How many faces does a triangular prism have?
8. How many edges does a triangular prism have?
9. How many vertices does a triangular prism have?
10. How many faces does a hexagonal pyramid have?
11. How many edges does a hexagonal pyramid have?
12. How many vertices does a hexagonal pyramid have?
13. How many faces does a rectangular prism have?
14. How many edges does a rectangular prism have?
15. How many vertices does a rectangular prism have?
16. How many faces does a triangular pyramid have?
17. How many edges does a triangular pyramid have?
18. How many vertices does a triangular pyramid have?

APPLICATIONS ***Name examples of each three-dimensional figure.***

19. square pyramid

20. cone

21. cylinder

22. sphere

SKILL 53

Name ______________________ Date __________

Surface Area of Rectangular Prisms

The **surface area** of a rectangular prism is equal to the sum of the areas of its faces.

EXAMPLE ***Find the surface area of the rectangular prism.***

Find the area of each face.

Front	$5 \times 10 = 50$
Back	$5 \times 10 = 50$
Top	$5 \times 4 = 20$
Bottom	$5 \times 4 = 20$
Right Side	$4 \times 10 = 40$
Left Side	$4 \times 10 = 40$

10 in.
4 in.
5 in.

Add the areas.

$50 + 50 + 20 + 20 + 40 + 40 = 220$

The surface area of the rectangular prism is 220 square inches.

EXERCISES ***Find the surface area of each rectangular prism.***

1.

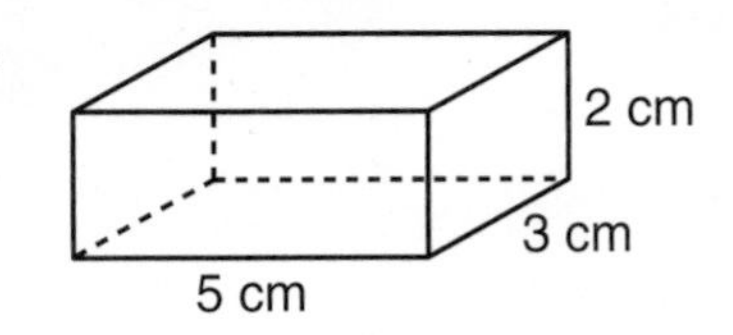

2.

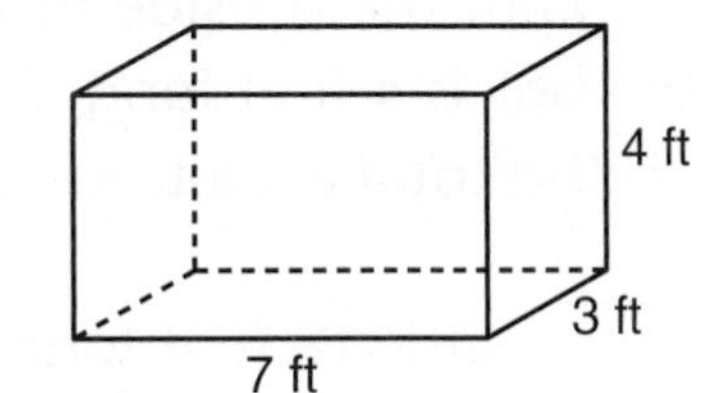

3.

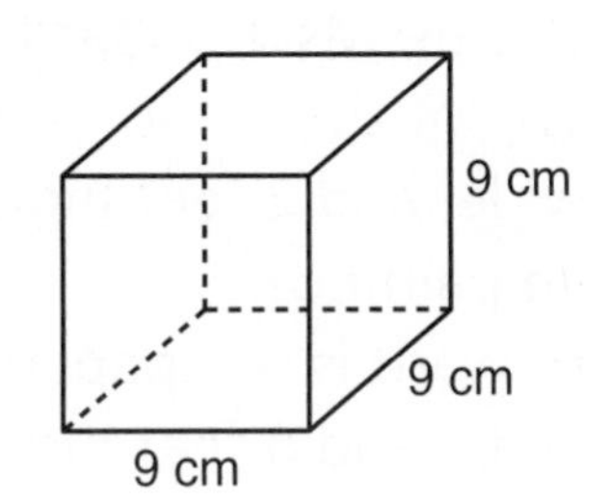

4.

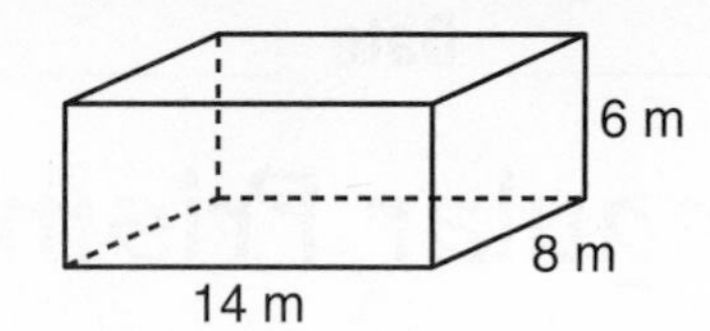

5.

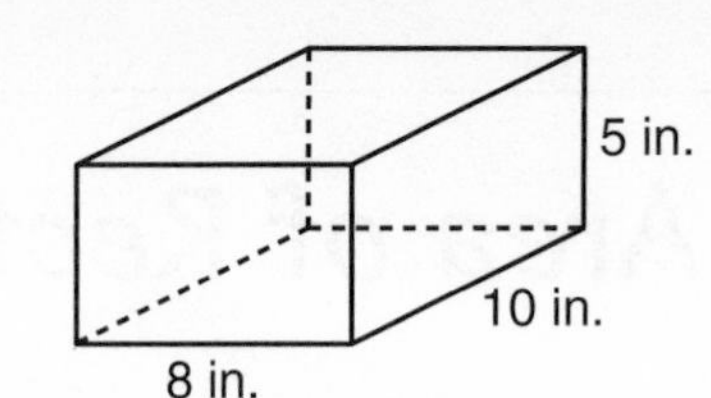

6.

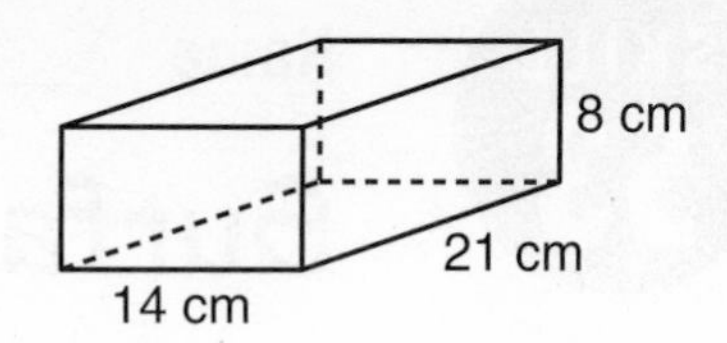

7.

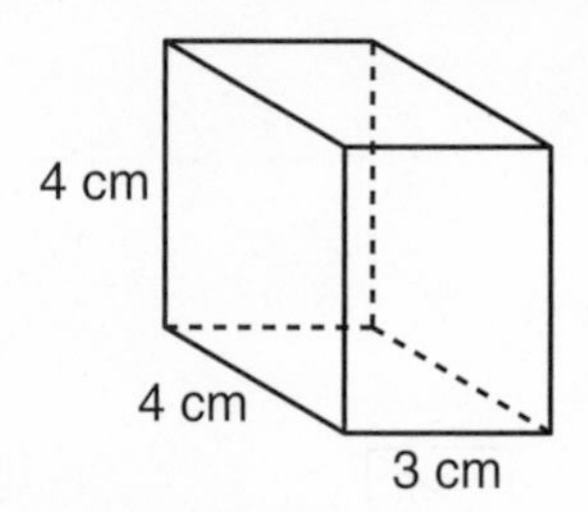

8.

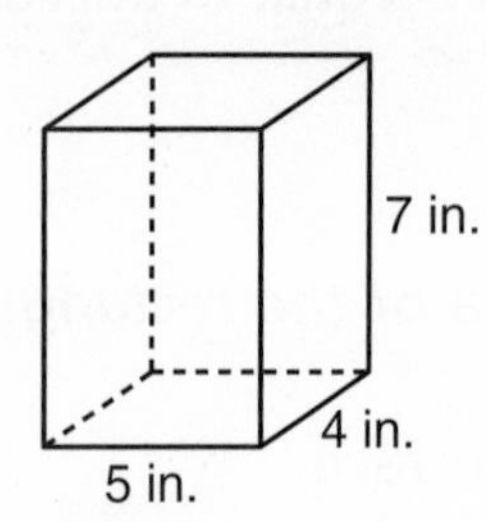

9.

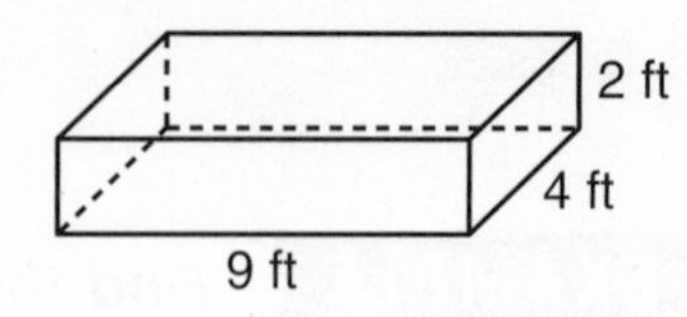

10.

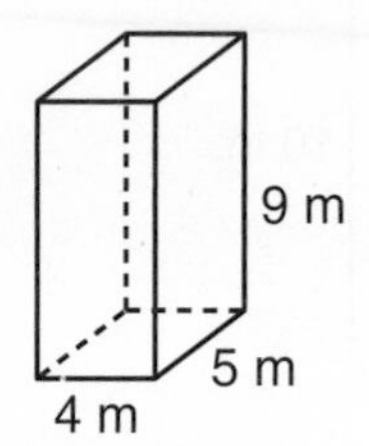

11.

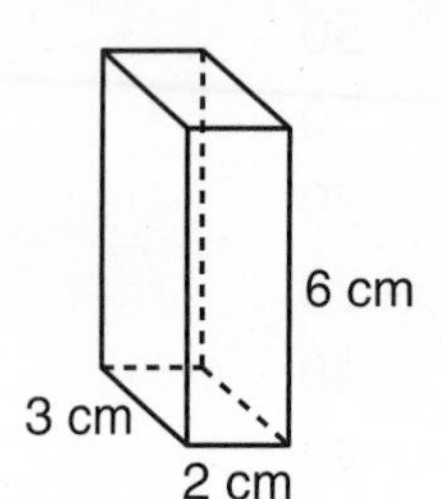

12.

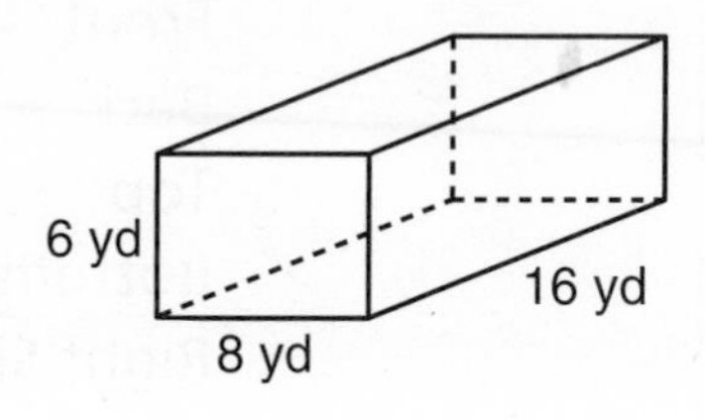

APPLICATIONS

13. A cereal box is 19 centimeters long, 6 centimeters wide, and 28 centimeters high. An artist is trying to create a design for the box. What is the surface area the artist needs to cover?

14. Jackson Middle School has a large storage box that is used for storing balls and other supplies for the physical education classes. Ms. Hubbard wants to paint the outside of the box one of the school colors. If the box is 4 feet long, 3 feet wide, and 2 feet high, what is the total area that needs to be painted?

15. Howard is wallpapering a room that is 18 feet long, 14 feet wide, and 8 feet high. How much wallpaper is needed to cover the walls, not taking into account doorways or windows?

16. One gallon of paint covers about 400 square feet of wall. If paint costs $17.99 a gallon and the sales tax is 5%, how much will it cost to put two coats of paint on the walls of your classroom?

SKILL 54

Name ______________________ Date __________

Surface Area of Cylinders

To find the surface area of a cylinder, find the sum of the areas of the two circular bases and the area of the rectangular face.

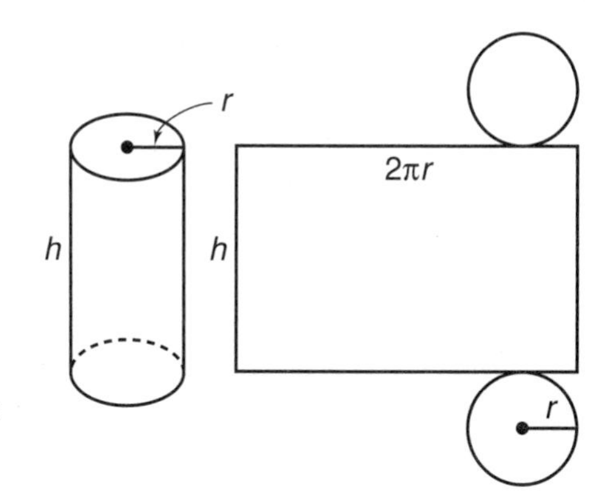

$$\text{area of each circle} = \pi r^2$$
$$\text{area of both circles} = 2\pi r^2$$

The length of the rectangle is equal to the circumference of the circle. So the area of the rectangle is $2\pi r \times h$.

The surface area of a cylinder is $2\pi r^2 + 2\pi rh$.

EXAMPLE

Find the area of the cylinder shown at the right.

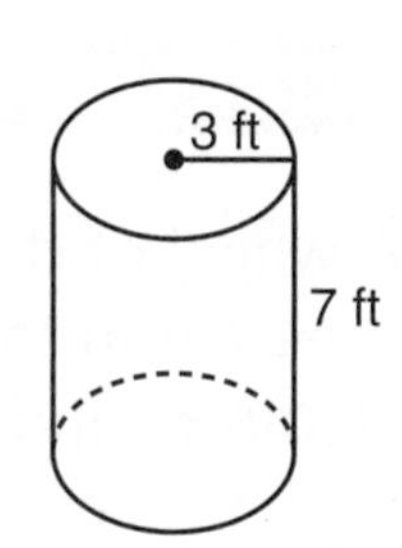

$$A = 2\pi r^2 + 2\pi rh$$
$$A \approx (2 \times 3.14 \times 3^2) + (2 \times 3.14 \times 3 \times 7)$$
$$A \approx 56.52 + 131.88$$
$$A \approx 188.40$$

The surface area of the cylinder is 188.40 square feet.

EXERCISES

Find the surface area of each cylinder. Use 3.14 for π.

1.

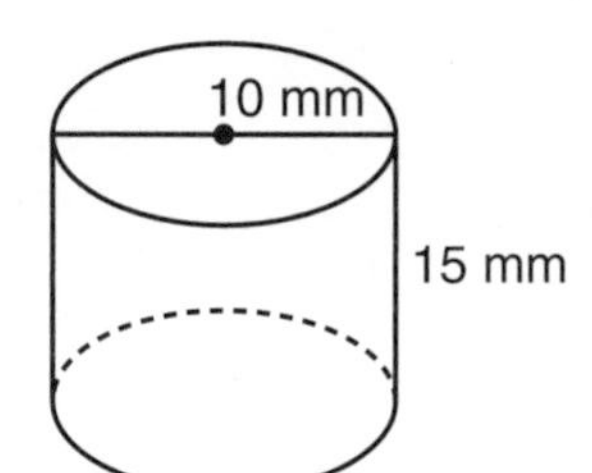

2.

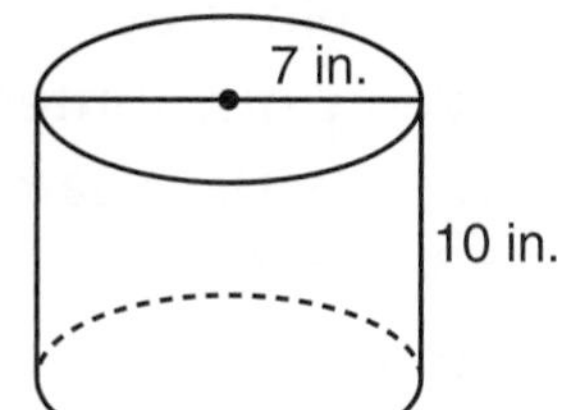

3.

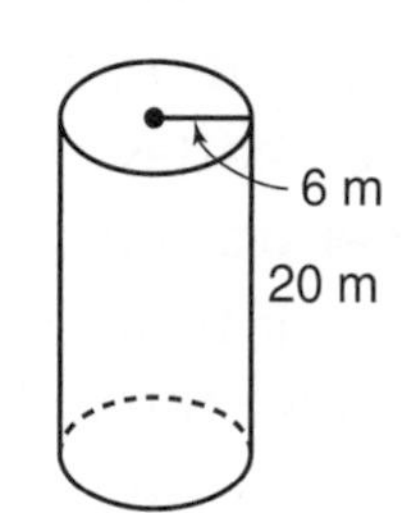

4.

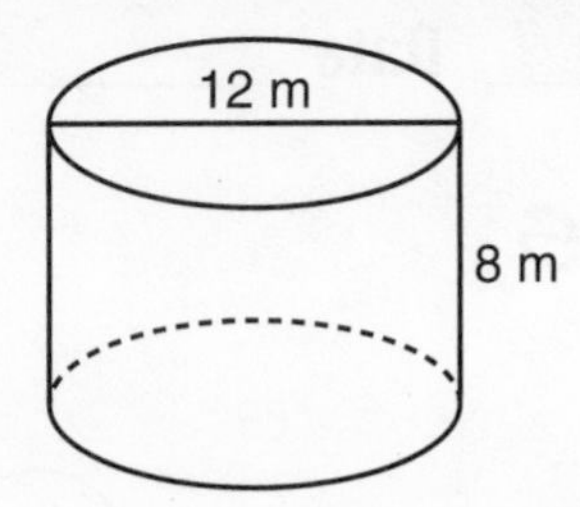

5.

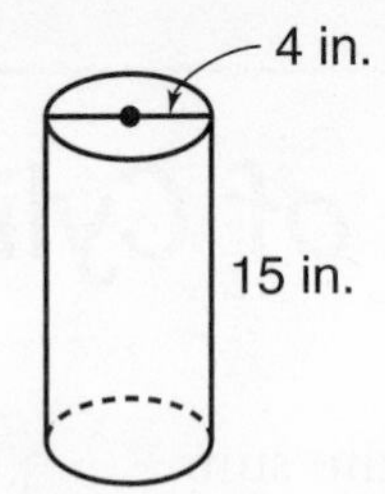

6.

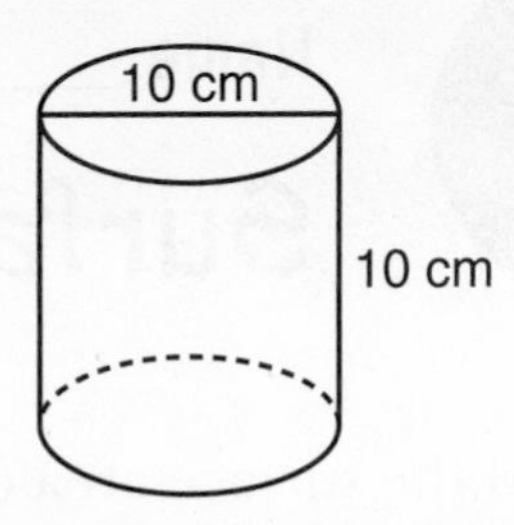

7.

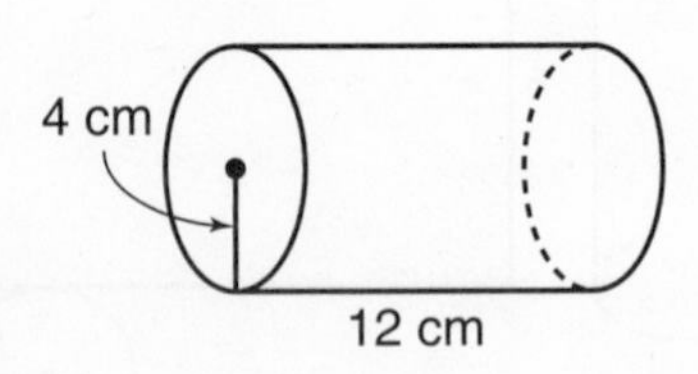

8.

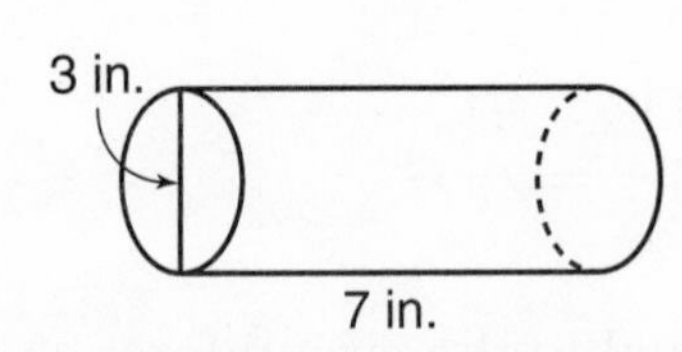

9.

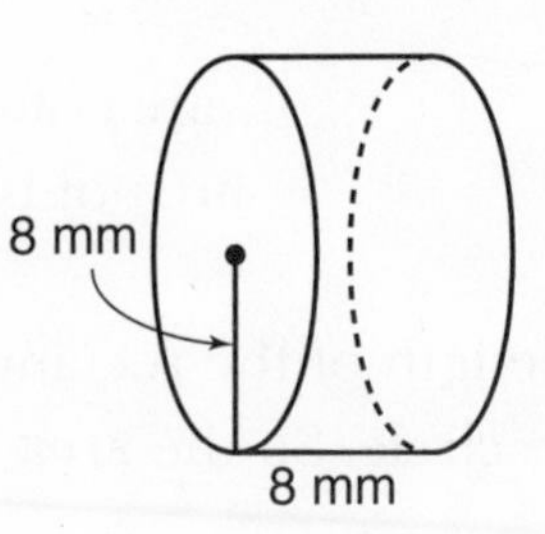

APPLICATIONS

10. A wheel of cheese is sealed in a wax covering. The wheel of cheese is in the shape of a cylinder that has a diameter of 10 inches and a height of 5 inches. What is the surface area of the cheese that needs to be covered in wax?

11. A storage tank is in the shape of a cylinder that has a radius of 2 feet and a height of 8 feet. The tank needs to be painted. What is the surface area of the tank?

12. A biologist is conducting an experiment to determine the amount of beetle infestation in the bark of elm trees. She believes that the beetles are fairly evenly distributed throughout the lower portions of the tree's bark. A sample of one square foot of bark from one tree two feet in diameter had 20 beetles. Consider the tree trunk to be a cylinder and determine how many beetles are probably in the first 10 feet of the tree's bark.

SKILL 55

Name ______________________________ Date ____________

Volume of Rectangular Prisms

The **volume** of an object is the amount of space that a solid contains. Volume is measured in cubic units. The volume (V) of a rectangular prism is equal to the product of the length (ℓ) times the width (w) times the height (h).

$$V = \ell wh$$

EXAMPLE ***Find the volume of the rectangular prism at the right.***

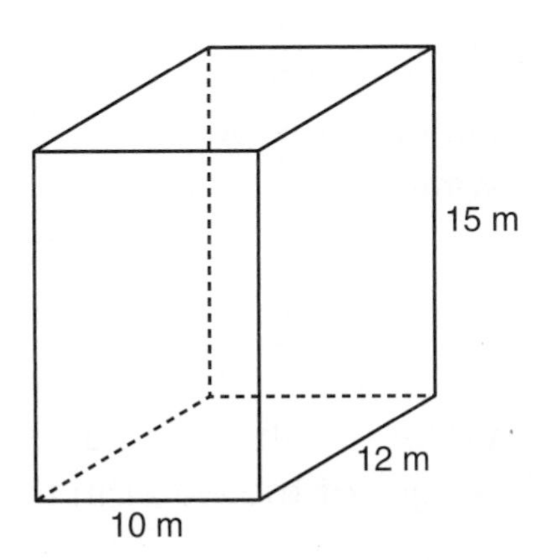

The length of the rectangular prism is 10 meters, the width is 12 meters, and the height is 15 meters.

$V = \ell wh$

$V = 10 \times 12 \times 15$

$V = 1{,}800$

The volume of the rectangular prism is 1,800 cubic meters.

EXERCISES ***Find the volume of each rectangular prism.***

1.

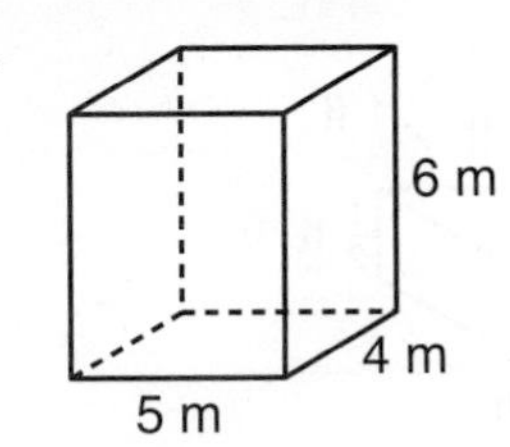

2.

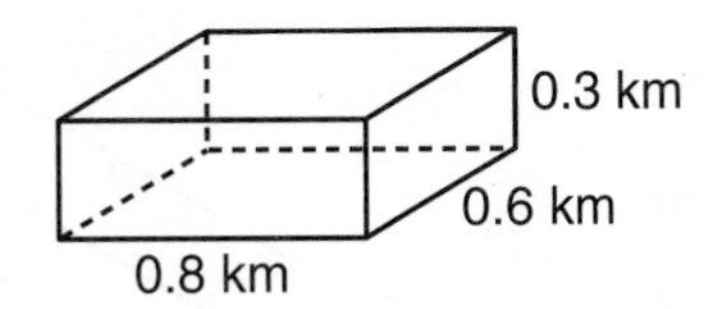

3.

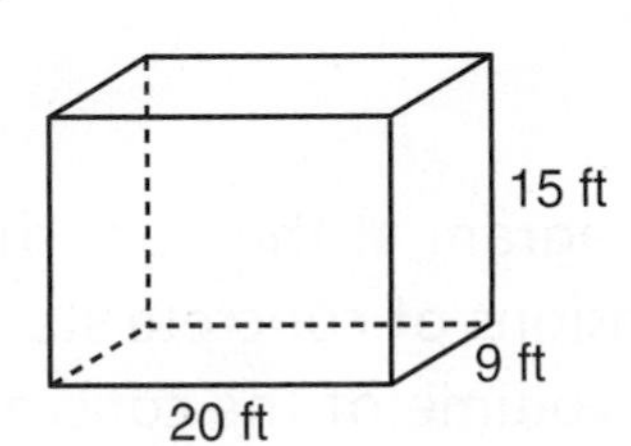

4.

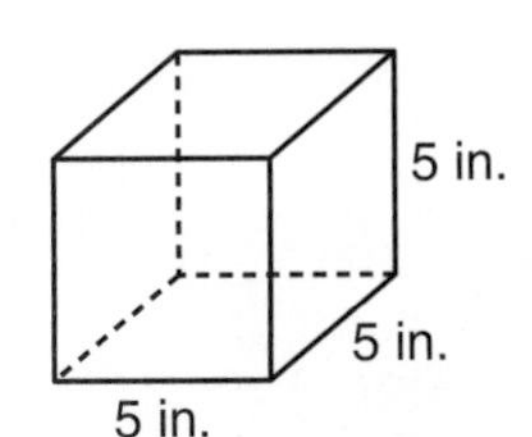

5.

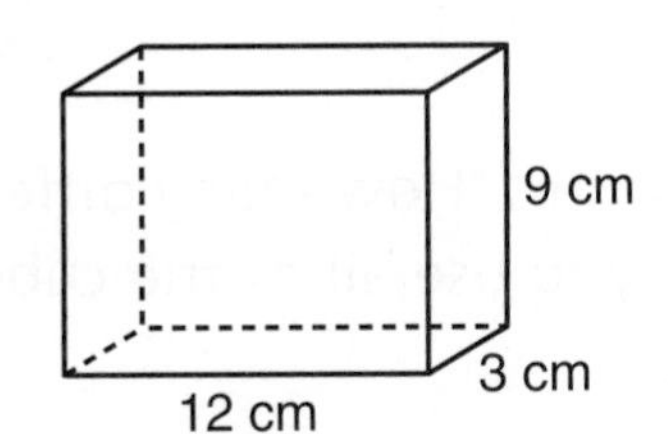

6.

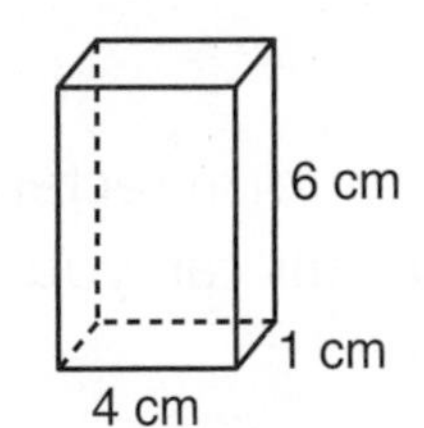

7.

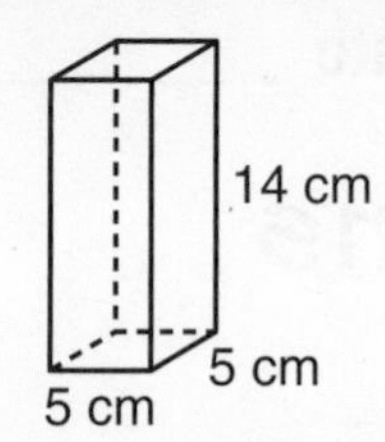

8.

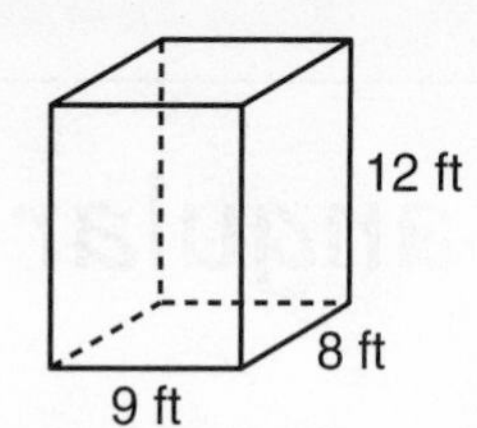

9.

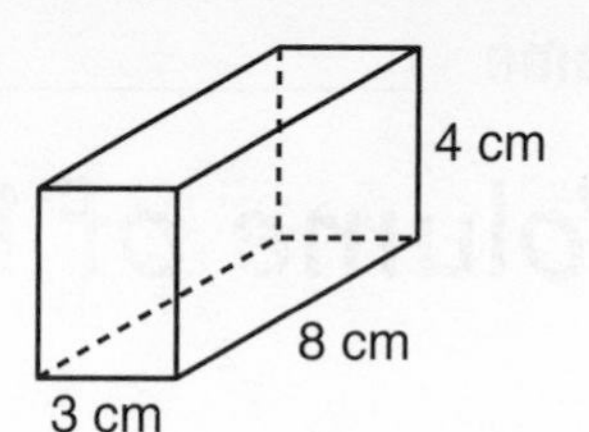

APPLICATIONS

10. The Pomerantz family have a small rectangular pond in their flower garden. The pond is 6 feet long and 4 feet wide. If the water in the pond is 2 feet deep, what is the volume of the water?

11. Water weighs about 62 pounds per cubic foot. What is the weight of the water in the pond in Exercise 10?

12. Janine keeps her jewelry in a jewelry box that measures 9 inches by 4.5 inches by 3 inches. What is the volume of the jewelry box?

13. The diagram at the right shows the dimensions of concrete stairs. What is the volume of the concrete?

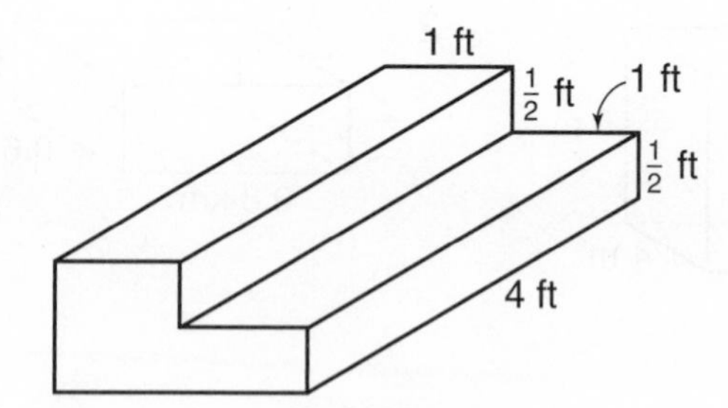

14. Use 20 cubes to form rectangular prisms. How many different rectangular prisms can you make if you use all of the cubes for each prism?

Name ______________________ Date ____________

Volume of Cylinders

The volume of a cylinder is found by multiplying the area of the base (πr^2) times the height (h).

$$V = \pi r^2 h$$

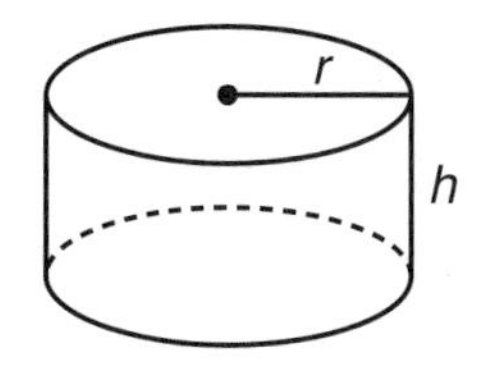

EXAMPLE ***Find the volume of the cylinder at the right.***

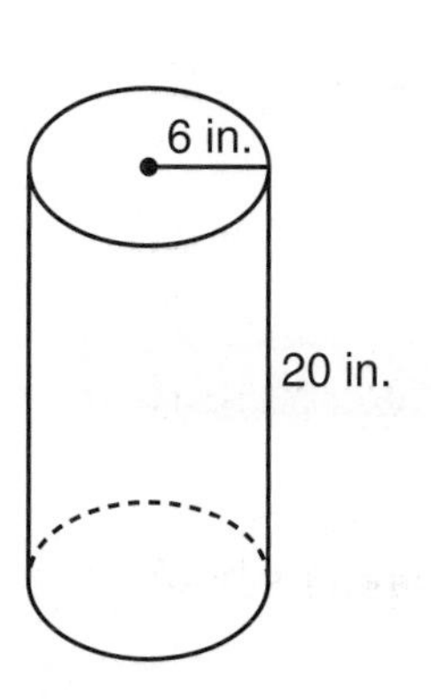

This cylinder has a radius of 6 inches and a height of 20 inches.

$V = \pi r^2 h$
$V = \pi(6)^2(20)$
$V = \pi(36)(20)$
$V \approx 3.14(36)(20)$ *Use 3.14 for π.*
$V \approx 2{,}260.8$

The volume of the cylinder is about 2,260.8 cubic inches.

EXERCISES ***Find the volume of each cylinder. Use 3.14 for π.***

1. 8 ft, 16 ft

2. 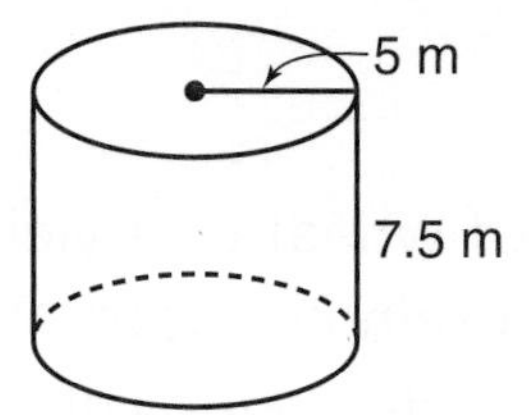

3. 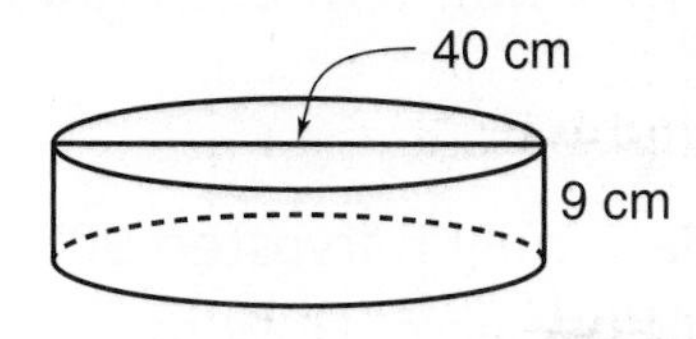

4.

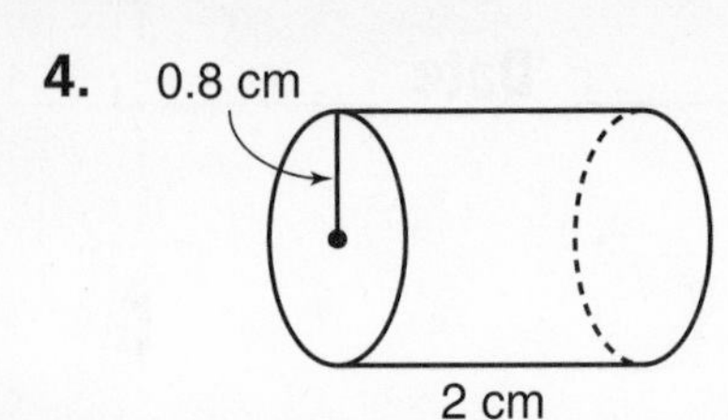

5.

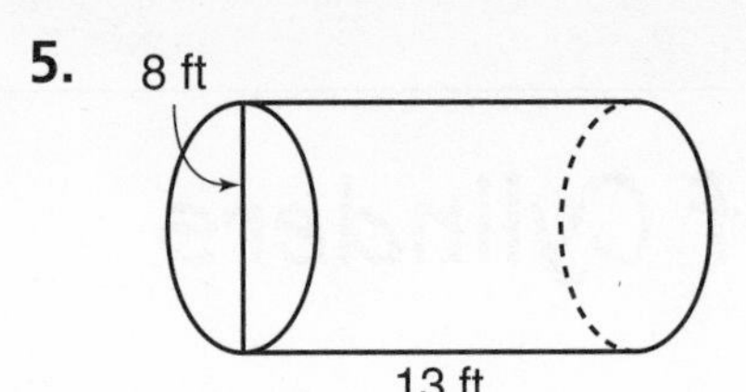

6.

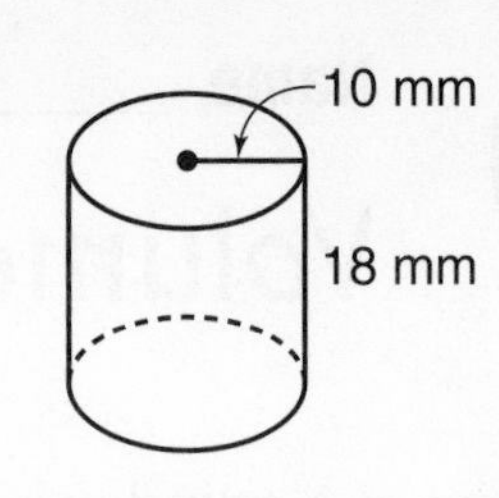

7.

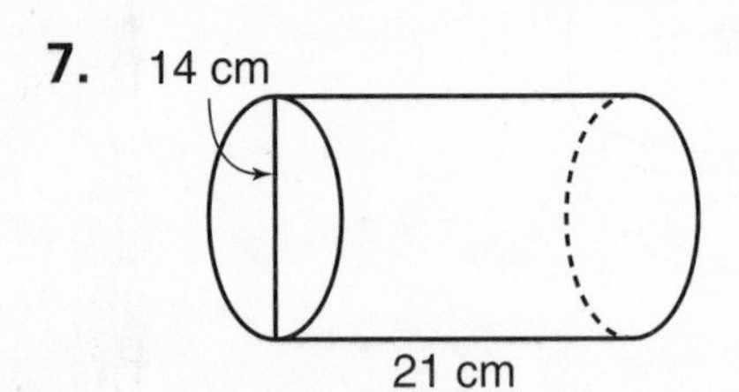

8.

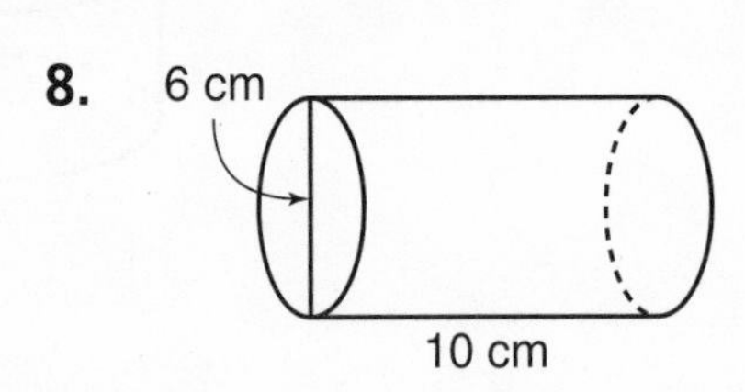

9.

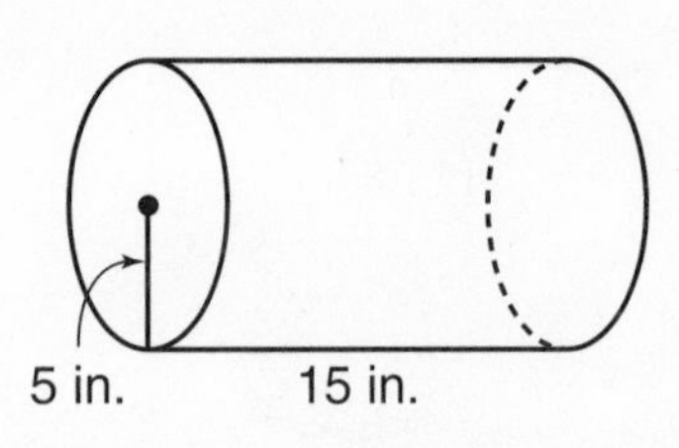

APPLICATIONS

10. A cylindrical water tank on the Okida family farm is 30 feet in diameter and 20 feet high. Find the volume of the tank.

11. The Okida family has 250 cows, and each cow drinks about 8 gallons of water per day. How many days will the tank in Exercise 10 provide the cows with water? (Hint: One cubic foot is about 7.5 gallons.)

12. The Okida family stores their wheat in cylindrical grain elevators that are each 20 feet in diameter and 60 feet high. What is the volume of each grain elevator?

13. The Okida family harvested 900 acres of wheat that yielded 40 bushels per acre. How many of the grain elevators in Exercise 12 are needed for the harvest? (Hint: One cubic foot of wheat is about 0.8 bushels.)

14. A coffee can is 6.5 inches high and has a diameter of 5 inches. Find the volume of the can.

Name ______________________________ Date __________

Measuring Angles

To measure an angle, place the center of a protractor on the vertex of the angle. Place the zero mark of the scale along one side of the angle. Read the angle measure where the other side of the angle crosses the scale.

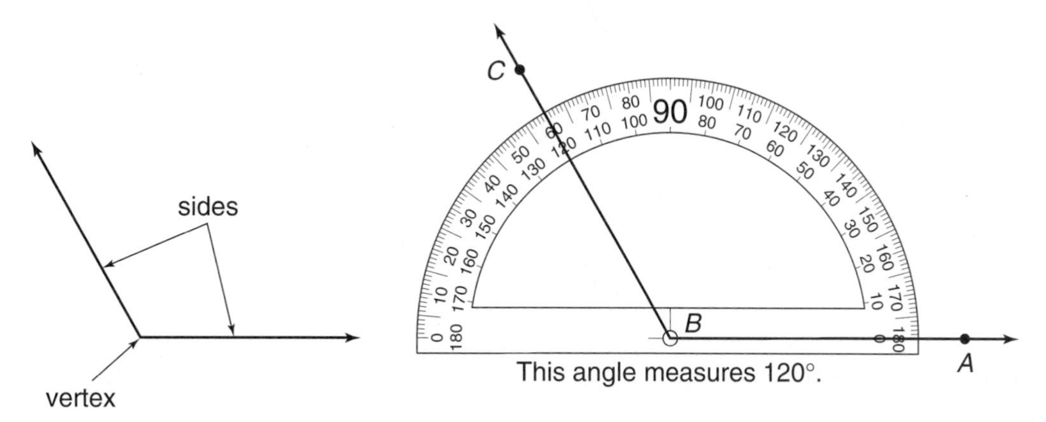

This angle measures 120°.

Angles may be classified according to their measure.

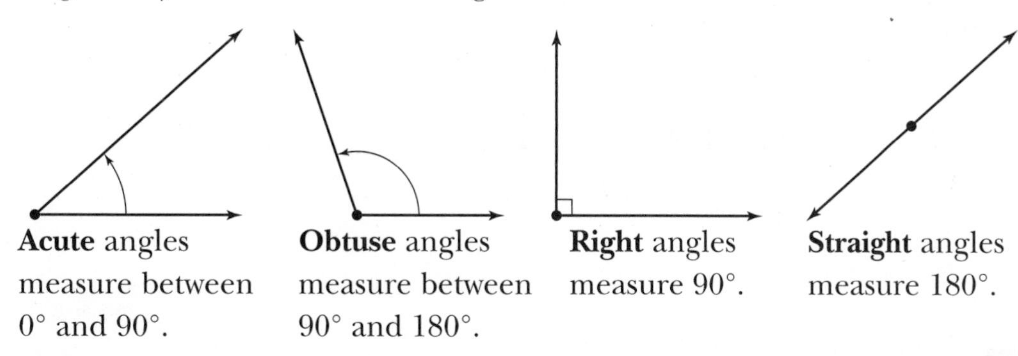

Acute angles measure between 0° and 90°.

Obtuse angles measure between 90° and 180°.

Right angles measure 90°.

Straight angles measure 180°.

EXERCISES *Classify each angle as acute, right, or obtuse.*

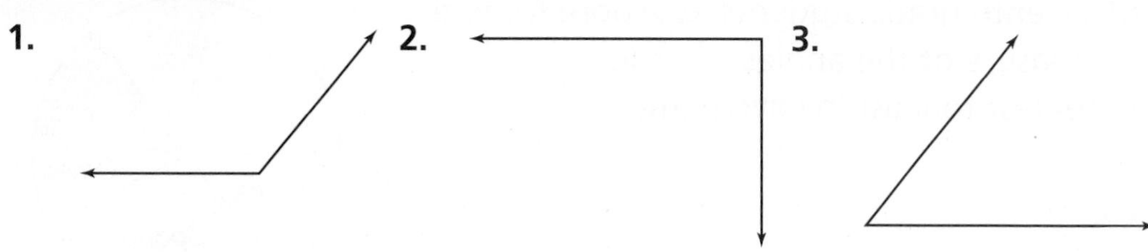

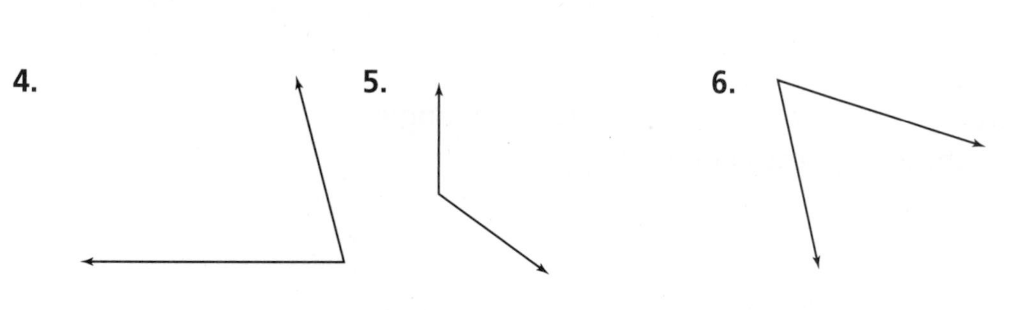

Use a protractor to find the measure of each angle.

7.

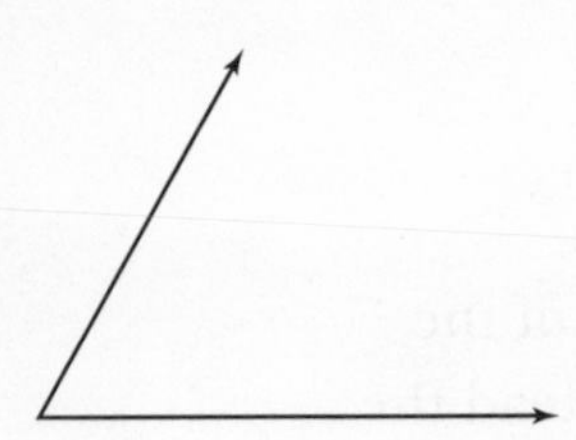

8. **9.**

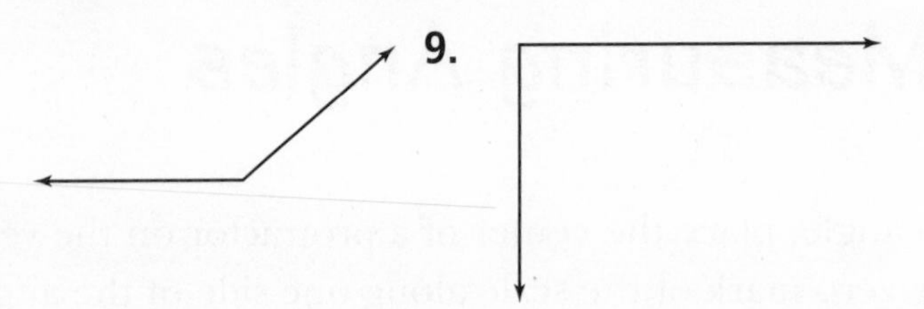

10.

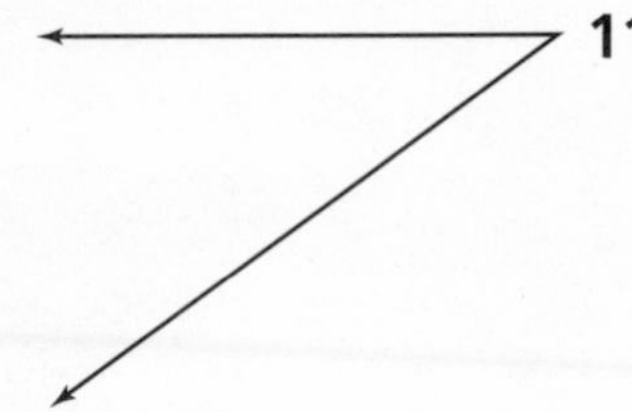

11.

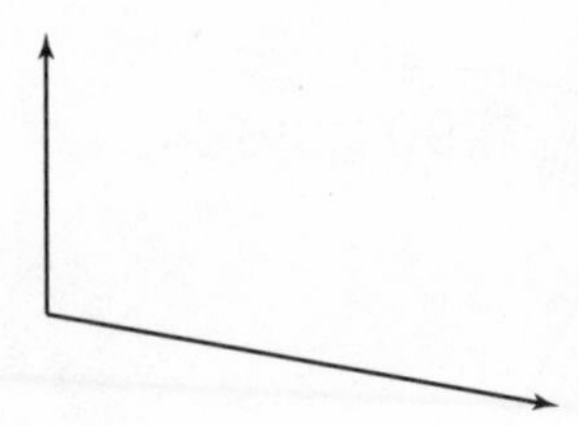

12.

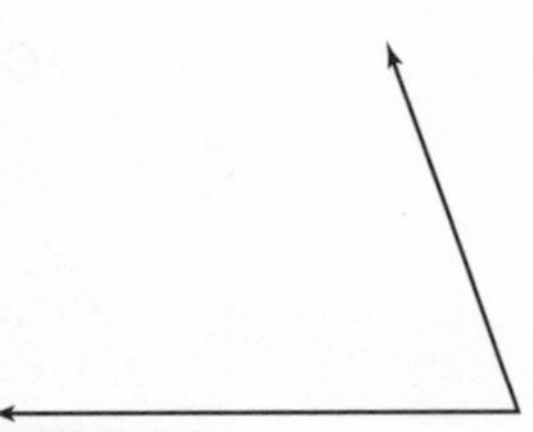

Classify angles having each measure as acute, right, obtuse, or straight.

13. 47° **14.** 95° **15.** 180° **16.** 82.9°

17. 90° **18.** 153° **19.** 179° **20.** 25°

APPLICATIONS

21. The circle graph at the right shows the after-school participation of seventh grade students at Moore Middle School. Use the measure of the angles to order the activities from greatest to least involvement.

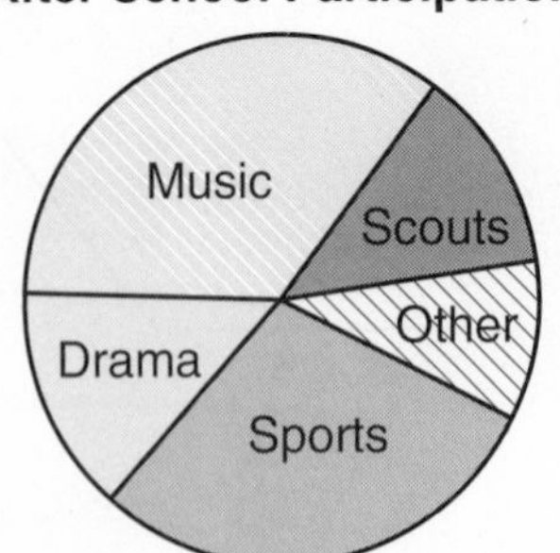

22. Without a protractor, draw your best estimate of an angle measuring 105°. Check your estimate with a protractor.

Name ______________________ Date __________

Metric Units of Measure

The **meter** is the basic unit of length in the metric system. Other metric units of length are **millimeters**, **centimeters**, and **kilometers**. Metric units of length are related in the following ways:

1 millimeter (mm) = 0.001 meter (m)
1 centimeter (cm) = 0.01 meter
1 kilometer (km) = 1,000 meters

EXAMPLE ***The metric ruler shown below can be used to measure the length of the paper clip in centimeters and in millimeters.***

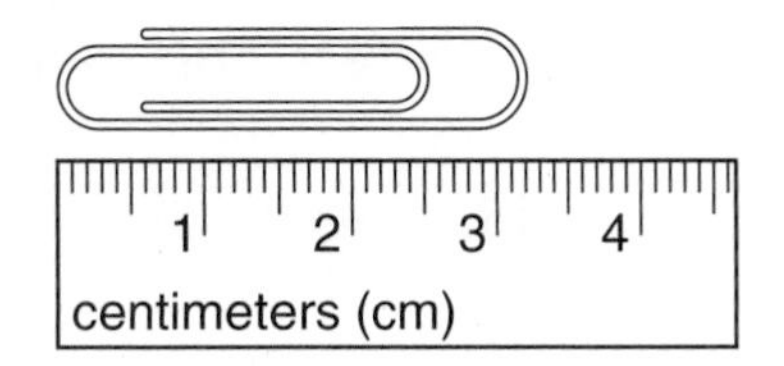

Centimeters:
The distance between two numbered marks is a centimeter. Each centimeter is divided into tenths. Therefore, the paper clip is about 3.2 centimeters long.

Millimeters:
The distance between two smaller marks is a millimeter. There are 10 millimeters in one centimeter. Therefore, the paper clip is about 32 millimeters long.

EXERCISES ***Complete each sentence with the most reasonable unit. Write* millimeters, centimeters, meters, *or* kilometers.**

1. Jack rode 5 ______________ along the bike trail.
2. The length of the room is about 6 ______________ .
3. The width of a kite is about 85 ______________ .
4. The button from your coat is about 5 ______________ thick.

Circle the best estimate.

5.	length of a river	500 cm	500 m	500 km
6.	length of a quilt	2.5 cm	2.5 m	2.5 km
7.	length of a cassette tape	10 mm	10 cm	10 m
8.	thickness of a rope	9 mm	9 cm	9 m
9.	diameter of a bicycle wheel	65 cm	65 m	65 km
10.	length of a bolt	20 cm	20 m	20 mm
11.	length of a bus route	15 cm	15 m	15 km

Find the length of each object in centimeters and millimeters.

12.

13.

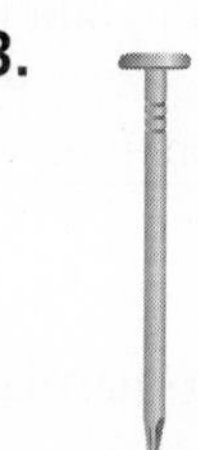

14.

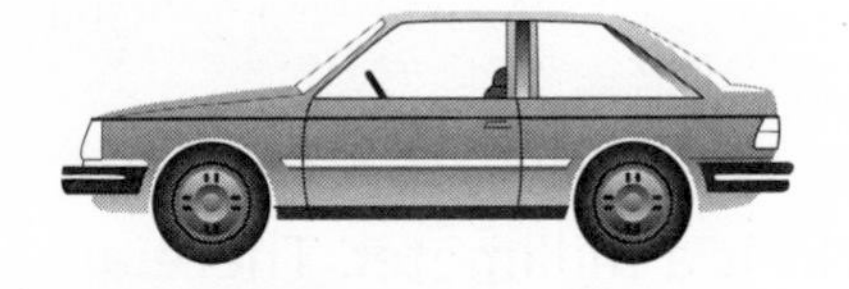

15.

APPLICATIONS

16. Name three objects whose lengths are between 1 centimeter and 1 meter.

17. Estimate the length of your pencil in centimeters. Then measure to check your estimate.

18. Juan ran 0.5 kilometers and Maria ran 600 meters. Who ran farther?

19. Mrs. Miller bought 2.75 meters of blue ribbon and 3.75 meters of red ribbon. How many meters of ribbon did she buy in all?

Name ______________________ Date __________

Customary Units of Measure

The **inch** is a customary unit of length used in the United States. Other customary units of length are **feet**, **yards**, and **miles**. Customary units of length are related in the following ways:

1 foot (ft) = 12 inches (in.)
1 yard (yd) = 36 inches or 3 feet
1 mile (mi) = 5,280 feet or 1,760 yards

EXAMPLES ***Which is the best estimate for the width of a kitchen table: 4 inches, 4 feet, 4 yards, or 4 miles?***

The width of a kitchen table is much greater than 4 inches, and much less than 4 miles. So, it is either 4 feet or 4 yards. The length of a 3-ring binder is an estimate for 1 foot. About four 3-ring binders would fit across the table. Therefore, the best estimate is 4 feet.

Find the length of the toothpick to the nearest eighth inch.

Line up the toothpick with 0 on a ruler. The toothpick is $2\frac{5}{8}$ inches long.

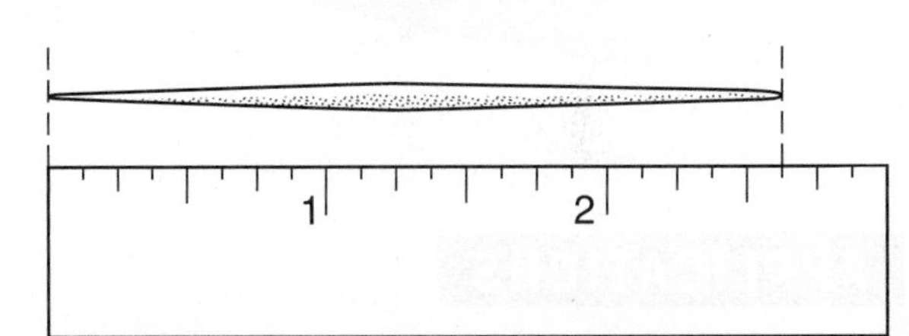

EXERCISES ***Complete each sentence with the most reasonable unit. Write* inches, feet, yards, *or* miles.**

1. Chen walked 3 ______________ along the hiking path.
2. The length of a bed is 6 ______________ .
3. South Carolina is about 285 ______________ .
4. The height of your classroom is about 3 ______________ .

Circle the best estimate.

5. length of a goldfish	2 in.	2 ft	2 yd
6. width of a driveway	14 in.	14 ft	14 yd
7. height of an elephant	3 in.	3 yd	3 mi
8. width of a TV screen	15 in.	15 ft	15 mi

Find the length of each object to the nearest eighth inch.

9.

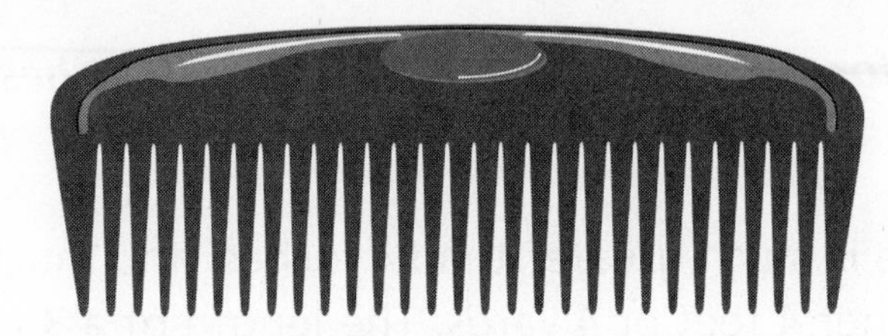

10.

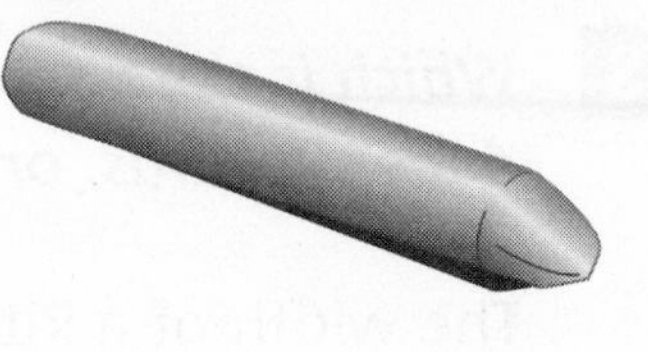

11.

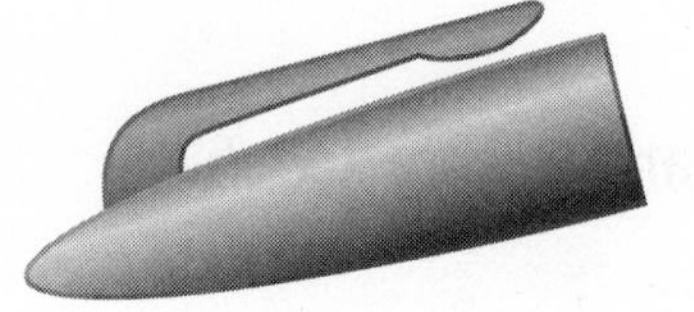

12.

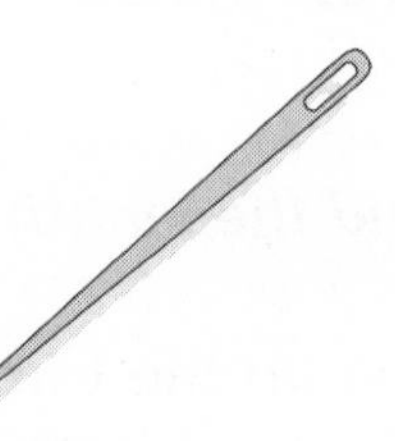

APPLICATIONS

13. Pat is 5 feet tall, and Kay is 58 inches tall. Which of them is taller?

14. Name three objects whose lengths are between 1 foot and 1 yard.

15. Mark an ✘ in the box that names the best unit for measuring each item. You may choose more than one unit.

Item	inches	feet	yards	miles
height of a student				
width of a book				
distance between cities				
length of a race				

Name ______________________________ Date ____________

Length

Commonly used units of length are listed below.

Customary System
1 foot = 12 inches 1 yard = 36 inches or 3 feet 1 mile = 5,280 feet or 1,760 yards

Metric System
1 centimeter = 10 millimeters 1 meter = 100 centimeters 1 kilometer = 1,000 meters

EXAMPLE ***What is the length of the large size paper clip shown below in customary units and metric units?***

Since the paper clip is not a long item, use inches and centimeters to measure it.

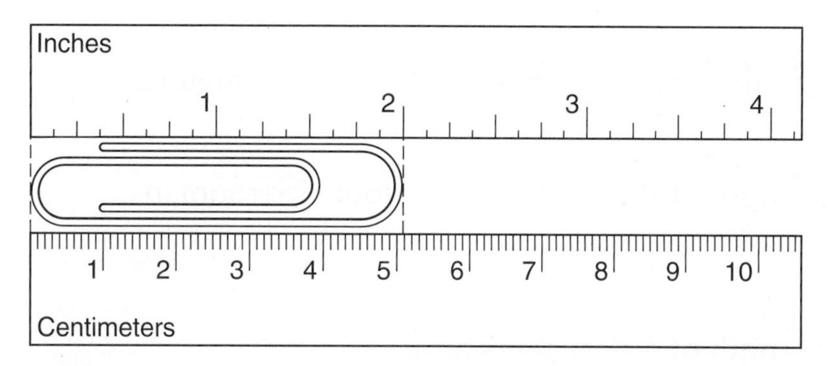

The paper clip is 2 inches or 5.1 centimeters long.

EXERCISES ***Measure each line segment to the nearest eighth inch.***

1. ____

2. ___________________________

3. ____

4. ___________

5. ___________________________________

Measure each line segment to the nearest tenth centimeter.

6. ____________________

7. ___________

8. ____

9. ______________________________

10. ____

APPLICATIONS

11. Measure the length of your pen in centimeters.

12. Measure the height of the seat of your chair in inches.

13. Measure the length and width of your classroom in meters.

14. Measure the length of the chalkboard in your classroom in feet.

15. Measure the length of a classmate's arm.

16. Measure the height of your classroom door.

17. Measure the length and width of your bedroom.

18. Measure the height of your kitchen table.

19. Measure the length of a spoon.

20. Measure the length of your comb.

Name ______________________________ Date ____________

Precision and Significant Digits

The piece of square tile shown at the right has sides that measure 0.050 m.

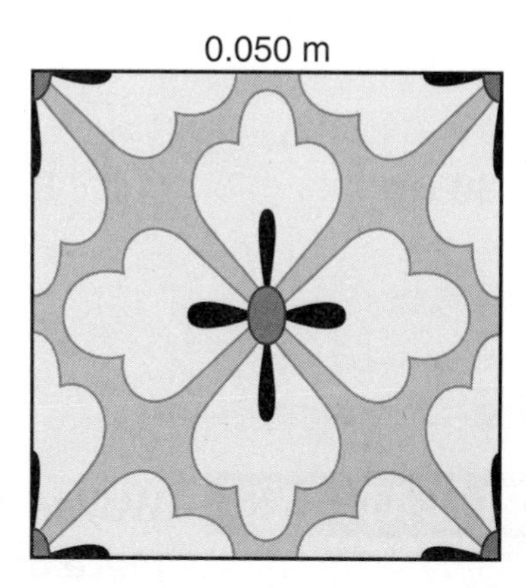

EXAMPLE ***Give the precision of this measurement. Then give the measurements between which the actual length lies and the number of significant digits.***

The precision of this measurement is the unit of measure, which is 0.001 m.

The measurements between which the actual length lies are 0.049 m and 0.051 m.

There are two significant digits. The zeros before the 5 are used only to show the place value of the decimal and are not counted as significant digits. However, the zero after the 5 is significant.

EXERCISES ***Determine the number of significant digits in each measure.***

1. 137 cm

2. 44 mm

3. 58 m

4. 6.0 cm

5. 3.25 in.

6. 5.60 ft

7. 19.08 m

8. 0.7 mm

9. 0.006 m

10. 2.70 yd

11. 50 ft

12. 2.010 m

13. 9.30 cm

14. 8.05 in.

15. 10.30 m

APPLICATIONS ***Wayne and Nora both measure the same piece of copper tubing. Wayne says the length is 14.2 cm. Nora says the length is 142.1 mm. Use this information to answer Exercises 16–18.***

16. Which measurement is more precise? Why?

17. Which measurement has the greater number of significant digits? Does this make this measurement better than the other? Why or why not?

18. Do you think either Wayne or Nora measured inaccurately? Why or why not?

19. Describe a situation in which precision of measurement would be very important. Describe a situation in which precision would not be very important.

Name ______________________ Date __________

Measuring Mass Using Customary Units

The most commonly used customary units of weight are **ounce**, **pound**, and **ton**.

1 pound (lb) = 16 ounces (oz)
1 ton (T) = 2,000 pounds

EXAMPLE ***A small car weighs 1.2 tons. How many pounds does the car weigh?***

Since there are 2,000 pounds in a ton, multiply 1.2 by 2,000 to find the weight in pounds.

1.2 tons = □ lb
1.2 tons = 1.2 × 2,000 lb
1.2 tons = 2,400 lb

The car weighs 2,400 pounds.

EXERCISES ***Complete.***

1. 3 lb = ____ oz **2.** 4 tons = ____ lb **3.** 1.25 lb = ____ oz

4. 3.2 tons = ____ lb **5.** 6,000 lb = ____ tons **6.** 32 oz = ____ lb

7. 9,000 lb = ____ tons **8.** $\frac{1}{2}$ lb = ____ oz **9.** 0.75 tons = ____ lb

10. 56 oz = ____ lb **11.** 3.125 lb = ____ oz **12.** 8.05 tons = ____ lb

13. 70 oz = ____ lb

14. $5\frac{1}{4}$ lb = ____ oz

15. 5,800 lb = ____ tons

16. 18 oz = ____ lb ____ oz

17. 91 oz = ____ lb ____ oz

18. 7 lb 5 oz = ____ oz

19. 2 lb 7 oz = ____ oz

20. 133 oz = ____ lb ____ oz

APPLICATIONS ***Honey is sold in 3-pound jars. Use this information to answer Exercises 21–23.***

21. How many 12-ounce containers can be filled from one jar of honey?

22. How many 8-ounce containers can be filled from two jars of honey?

23. How many 14-ounce containers can be filled from three jars of honey? Will there be enoungh honey left over to fill an 8-ounce container?

24. The oil tanker Jahre Viking is the heaviest ship ever built. When fully loaded, it weighs 622,420 long tons. If one long ton is equal to 2,240 pounds, what is the weight of the ship in pounds?

25. Juanita made a batch of brownies. The brownies weigh $2\frac{1}{2}$ pounds. How many 2-ounce brownies can she cut from this batch?

26. Find the weight of a full container of one of your favorite foods. What part of the food in a full container do you usually eat in one serving? About how many ounces do you usually eat?

Name ______________________ Date __________

Measuring Mass Using Metric Units

The basic unit of mass in the metric system is the **gram (g)**. A **kilogram (kg)** is 1,000 grams. A **milligram (mg)** is 0.001 gram.

EXAMPLE ***Jacob is filling small boxes with raisins. Each small box will hold 50 grams of raisins. How many small boxes can he fill using a 4-kilogram bag of raisins?***

To answer this question, first change 4 kilograms to grams.

$$4 \text{ kg} = \square \text{ g}$$
$$4 \text{ kg} = 4 \times 1{,}000 \text{ g}$$
$$4 \text{ kg} = 4{,}000 \text{ g}$$

Then divide 4,000 grams by 50 grams to find the number of small boxes that can be filled.

$$4{,}000 \div 50 = 80$$

Jacob can fill 80 small boxes with raisins.

EXERCISES ***Complete.***

1. 5 kg = _____ g **2.** 16 g = _____ mg **3.** 120 g = _____ kg

4. 180 kg = _____ g **5.** 150 mg = _____ g **6.** 0.6 kg = _____ g

7. 500 mg = _____ g **8.** 0.005 g = _____ mg **9.** 0.7 g = _____ kg

10. 60,000 mg = _____ g 11. 5.6 g = _____ mg 12. 7,800 g = _____ kg

13. 1,500 mg = _____ g 14. 0.003 kg = _____ g 15. 2.002 kg = _____ g

16. 8,500 mg = _____ kg 17. 0.009 kg = _____ mg 18. 1.018 kg = _____ mg

APPLICATIONS ***The USDA term "Lean" is used on meat and poultry products. It means that, per serving of the product, there are less than 10 grams of fat and 95 milligrams of cholesterol. Use this information to answer Exercises 19–25.***

19. How many milligrams of fat are there per serving of a lean product?

20. How many kilograms of fat are there per serving of a lean product?

21. Do you think it is better to give the amount of fat in milligrams, grams, or kilograms? Why?

22. How many grams of cholesterol are there per serving of a lean product?

23. How many kilograms of cholesterol are there per serving of a lean product?

24. Do you think it is better to give the amount of cholesterol in milligrams, grams, or kilograms? Why?

25. Suppose that a package of meat contains 8 servings and is labeled "Lean." What is the maximum number of grams of cholesterol the meat can contain?

26. How many 30-gram servings are contained in a box of cereal that has a mass of 0.403 kilograms?

Name ______________________ Date __________

Units of Time

1 hour (h) = 60 minutes (min)
1 minute (min) = 60 seconds (s)

To add measures of time, add the seconds, add the minutes, and add the hours. Rename if necessary.

EXAMPLE ***Add 4 hours 25 minutes 40 seconds and 5 hours 30 minutes 25 seconds.***

  4 h 25 min 40 s
+ 5 h 30 min 25 s
  9 h 55 min 65 s = 9 h 56 min 5 s

Rename 65 s as 1 min 5 s.

EXERCISES ***Rename each of the following.***

1. 14 min 85 s = __________ min 25 s

2. 8 h 65 min = 9 h __________ min

3. 3 h 19 min 67 s = 3 h __________ min 7 s

4. 6 h 68 min 25 s = __________ h __________ min 25 s

5. 7 h 105 min 15 s = __________ h __________ min 15 s

6. 4 h 99 min 80 s = __________ h __________ min __________ s

7. 1 h 76 min 91 s = __________ h __________ min __________ s

8. 7 h 88 min 60 s = __________ h __________ min __________ s

Add. Rename if necessary.

9. 35 min 45 s
+ 12 min 12 s

10. 6 h 50 min
+ 3 h 17 min

11. 9 h 45 min 10 s
+ 3 h 30 min 50 s

12. 1 h 55 min 12 s
+ 3 h 25 min 34 s

13. 11 h 33 min 6 s
+ 5 h 36 min 29 s

14. 6 h 10 min 47 s
+ 2 h 51 min 28 s

15. 7 h 30 min 52 s
+ 3 h 45 min 40 s

16. 9 h 10 min 45 s
+ 3 h 55 min 30 s

APPLICATIONS

An atlas gives average travel times. Use this information to answer Exercises 17–19.

Average Travel Times	
Baton Rouge to Mobile	4 h 40 min
Mobile to Tallahassee	5 h 50 min
Tallahassee to Jacksonville	3 h 35 min

17. What is the average travel time from Baton Rouge to Tallahassee going through Mobile?

18. What is the average travel time from Mobile to Jacksonville going through Tallahassee?

19. What is the average travel time from Baton Rouge to Jacksonville going through Mobile and Tallahassee?

20. Wesley Paul set an age group record in the 1977 New York Marathon. He ran the race in 3 hours 31 seconds. He was 8 years old at the time. If he ran 2 hours 58 minutes 48 seconds in practice the day before the race, for how long did Wesley run on both days?

Name ______________________ Date __________

Mean, Median, Mode

You can analyze a set of data by using three measures of center: mean, median, and mode.

EXAMPLE ***Hakeem Olajuwon, 1994's Most Valuable Player in the National Basketball Association, helped the Houston Rockets win the NBA championship. In winning the 7-game series, Olajuwon scored 28, 25, 21, 32, 27, 30, and 25 points. Find the mean, median, and mode of his scores.***

Mean: $\frac{28 + 25 + 21 + 32 + 27 + 30 + 25}{7} \approx 26.857$

The mean is about 27 points.

Median: 21, 25, 25, 27, 28, 30, 32

↑ median

The median is 27.

Mode: The mode is 25 since it is the number that appears the most times.

EXERCISES ***Find the mean, median, and mode for each set of data.***

1. 5, 4, 7, 2, 2, 1, 4, 3

2. 25, 18, 14, 27, 25, 16, 18, 25

3. 13, 11, 7, 9, 12, 5

4. 234, 163, 634, 267, 545, 874

5. 23, 36, 48, 95, 36, 28, 24

6. 299, 100, 237, 492, 333, 263, 295

7. 2,500, 2,366, 1,939, 1,933, 1,835, 2,498, 2,943

8. 9, 2, 5, 7, 8, 9, 4, 4, 6, 4

9. 29, 48, 20, 43, 33, 20, 40, 69, 48

10. 7,899, 4,395, 9,090, 9,588, 4,880, 9,587, 4,756

APPLICATIONS ***The data at the right shows the record high temperatures for several states in the U.S. Use the data to answer Exercises 11–15.***

State	Record High Temperature (°F)
Alabama	112
Alaska	100
Michigan	112
Oklahoma	120
Vermont	105
Wyoming	114

11. What is the mode?
12. What is the median?
13. What is the mean?
14. If each of the high temperatures increased by 1°F, would it change
    a. the mode? Why or why not?

    b. the median? Why or why not?

    c. the mean? Why or why not?

15. If the high temperature for Vermont increased to 112°F, would it change
    a. the mode? Why or why not?

    b. the median? Why or why not?

    c. the mean? Why or why not?

16. Find the hand spans of ten people. Ask each person to spread apart the little finger and thumb of his or her right hand as far as possible. Then measure and record the distance from tip to tip to the nearest centimeter. Find the mean, median, and mode for the data you collected.

Name ______________________ Date __________

Line Plots

Darrell surveyed some kennels to find the cost of grooming his dog. The prices are: \$25.00, \$27.00, \$32.00, \$22.00, \$43.00, \$28.00, \$18.00, \$24.00, \$25.00, \$27.00, \$30.00, \$24.00, \$22.00, \$30.00, \$12.00, \$25.00, and \$20.00.

EXAMPLE ***Organize this information using a line plot.***

The lowest price is \$12.00, and the highest price is \$43.00. Draw a number line that includes the numbers 12 to 43. Place an X above the number line to represent each price.

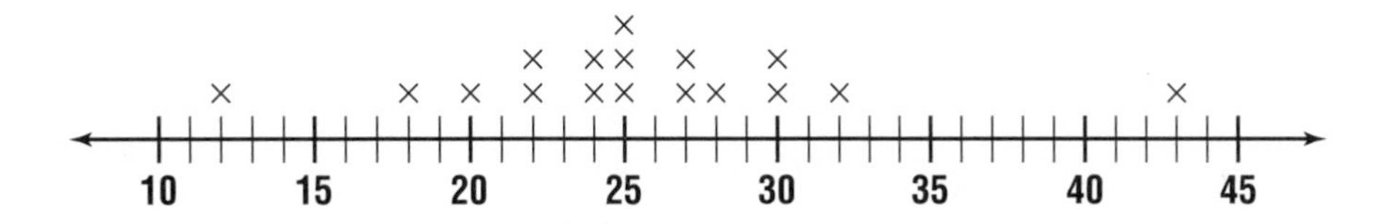

EXERCISES ***Make a line plot for each set of data.***

1. 23, 20, 23, 32, 35, 26, 35, 35, 44

2. 133, 139, 133, 139, 132, 132, 132, 132

3. 400, 600, 600, 200, 400, 1,000, 400

4. 5.3, 5.1, 5.0, 5.0, 6.0, 5.5, 5.3

5. 212, 215, 200, 203, 230, 227, 221, 218, 224

6. $4.30, $4.30, $4.10, $4.30, $4.30, $4.60, $4.10

APPLICATIONS ***Make a line plot for each set of data.***

7. The prices of appetizers at Ralph's Restaurant are listed below
$3.95, $6.95, $4.95, $3.95,
$5.95, $3.95, $4.95, $3.95,
$4.95, $4.95, $5.95, $4.95,
$3.95, $4.95, $4.95, $5.95

8. Miss Allen asked her students during which half-hour they usually wake up on school days. The results are listed below.
5:30, 7:00, 6:30, 6:00, 6:30, 7:00,
6:30, 6:30, 6:30, 7:00, 7:00, 6:30,
6:30, 7:30, 7:00, 6:30, 6:00, 7:00

9. The weights of the junior varsity wrestlers are listed below.
170, 160, 135, 135, 160, 122, 188, 154,
135, 140, 122, 103, 190, 154, 108, 150

10. The scores on a sixty-point history test are listed below.
55, 52, 49, 53, 38, 46, 52, 60, 55, 49,
32, 47, 55, 48, 60, 51, 47, 44, 37, 51

11. Ask your classmates to rate a certain television program from 1 to 10 with 10 being the best. Write down their responses and organize this information into a line plot.

Name ______________________________ Date __________

Stem-and-Leaf Plots

A **stem-and-leaf plot** is one way to organize a list of numbers. The **stems** represent the greatest place value in the numbers. The **leaves** represent the next place value.

EXAMPLE ***The fourteen states with the most representatives in the House of Representatives are listed below. Make a stem-and-leaf plot for this data.***

State	Representatives	State	Representatives
California	52	New Jersey	13
Florida	23	New York	31
Georgia	11	North Carolina	12
Illinois	20	Ohio	19
Indiana	10	Pennsylvania	21
Massachusetts	10	Texas	30
Michigan	16	Virginia	11

The stem will be the tens place and the leaves will be the ones place.

Stem	Leaves
1	00112369
2	013
3	01
4	
5	2

1|0 means 10 representatives.

EXERCISES ***Make a stem-and-leaf plot for each set of data.***

1. 56, 65, 57, 69, 58, 55, 52, 55, 66, 60, 53, 63

2. 230, 350, 260, 370, 240, 380, 290, 270, 220, 350, 300, 280

3. 4.5, 6.8, 5.2, 5.9, 5.1, 6.7, 4.0, 4.4, 6.0, 6.9

4. 1,900, 2,000, 2,600, 3,000, 2,500, 1,800, 2,200, 2,700, 1,600, 1,700, 2,000, 2,300

APPLICATIONS ***Each number below represents the age of workers at Fred's Fast Food.***

20 52 21 39 40 58 27 48 36 20 51 26
45 30 49 22 59 50 33 35 28 43 55 20

Use this data to answer Exercises 5–10.

5. Make a stem-and-leaf plot of the data.

6. How many people work at Fred's Fast Food?

7. What is the difference in the ages between the oldest and youngest workers at Fred's?

8. What is the most common age for a worker?

9. Which age group is most widely represented?

10. How many workers are older than 35 years?

11. Measure the length of your classmates' shoes in centimeters. Record the numbers and make a stem-and-leaf plot.

12. What is the most common length of your classmates' shoes?

Name ______________________ Date __________

Make a Table

The employees of Lake Products Corporation earn the following yearly salaries.

$14,500	$26,000	$43,200	$23,700	$33,400
$15,500	$28,900	$31,100	$56,300	$41,000
$35,000	$24,700	$16,300	$20,000	$63,000
$8,100	$22,800	$9,700	$32,200	$19,300

EXAMPLE ***Organize this information in a frequency table.***

Use intervals of $10,000 to make the frequency table.

Employee Salaries		
Salary	Tally	Frequency
$60,000–$69,999	|	1
$50,000–$59,999	|	1
$40,000–$49,999	||	2
$30,000–$39,999	||||	4
$20,000–$29,999	~~||||~~ |	6
$10,000–$19,999	||||	4
0–$9,999	||	2

EXERCISES ***Organize the information in a frequency table.***

1. Number of subscriptions sold on the first day of the club's fund-raising campaign:
   3, 0, 4, 2, 1, 0, 1, 1, 2, 4, 2, 3, 5, 0, 2, 1, 3, 1, 1, 2

APPLICATIONS

2. The Stanley Cup champions from 1977–1994:

1977 Montreal Canadiens
1978 Montreal Canadiens
1979 Montreal Canadiens
1980 New York Islanders
1981 New York Islanders
1982 New York Islanders
1983 New York Islanders
1984 Edmonton Oilers
1985 Edmonton Oilers
1986 Montreal Canadiens
1987 Edmonton Oilers
1988 Edmonton Oilers
1989 Calgary Flames
1990 Edmonton Oilers
1991 Pittsburgh Penguins
1992 Pittsburgh Penguins
1993 Montreal Canadiens
1994 New York Rangers

3. The results of your survey of your classmates' favorite movies

4. The results of your survey of your classmates' favorite pizza toppings

Name ______________________________ Date __________

Statistical Graphs

There are different types of statistical graphs. Three types of statistical graphs are bar graphs, circle graphs, and line graphs. A **bar graph** is used to compare quantities. A **circle graph** is used to compare parts to the whole. A **line graph** is used to show change.

EXAMPLE ***What type of graph should you use to compare the seating capacity of various aircraft?***

A bar graph is used to compare quantities, so a bar graph should be used. The graph at the right compares the capacity of the B747, the DC-10, the L-1011 and the MD-80. From the graph, it is easy to see that the B747 has the most seating capacity of these aircraft and the MD-80 has the least.

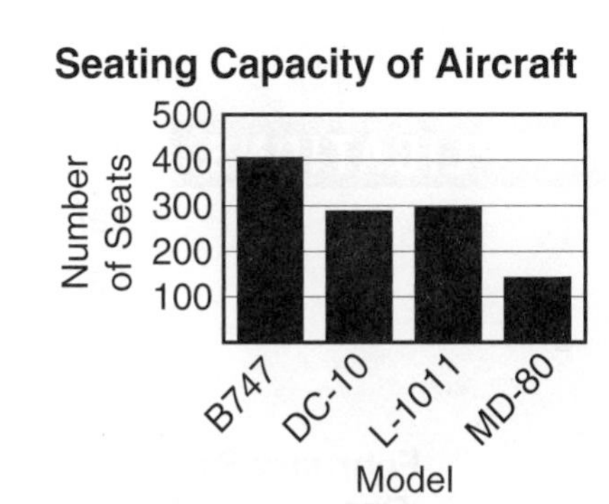

EXERCISES ***Determine whether you would use a bar graph, a circle graph, or a line graph to show the information.***

1. average temperature in Sacramento for each month of the year

2. average temperature in January of five California cities

3. land area of the continents

4. percentages of the total land area each continent represents

5. number of CD players sold each year from 1981 to 1994

6. weight of a baby in each month from birth to one year of age

7. percentages of sources of fuel in the United States

8. height of the five tallest buildings in the world

9. Zawodniak family budget

10. average weekly attendance at five different theaters

APPLICATIONS ***The following graphs show the weekly sales at Bennie's Bakery for the month of February. Use the graphs to answer Exercises 11–13.***

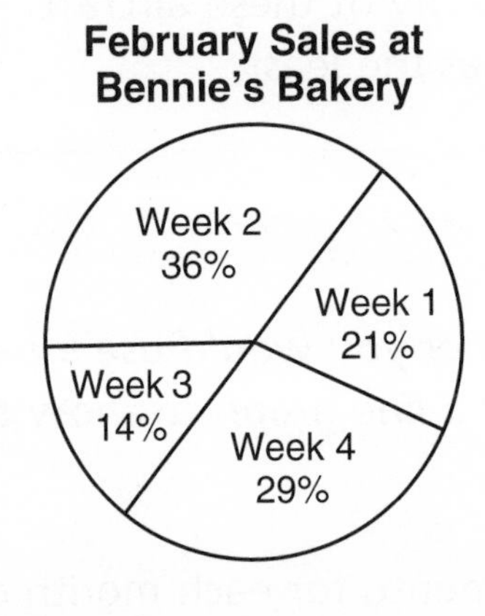

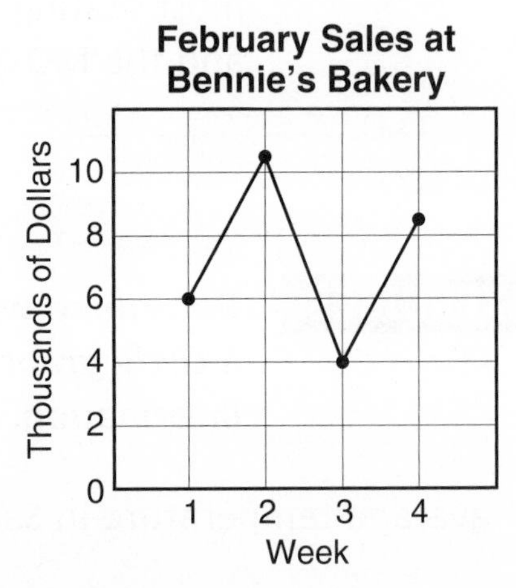

11. Which graph best shows how the sales for each week compare to each other?

12. Which graph best shows the changes in the sales over the four weeks?

13. Which graph best shows what part of February's sales each week represents?

Name ______________________ Date __________

Line Graphs

A **line graph** is usually used to show the change and direction of change over time. All line graphs should have a graph title, a vertical-axis label, and a horizontal-axis label.

EXAMPLE ***Make a line graph for the data on the number of space flights carrying people during the 1960's.***

Space Flights Carrying People	
Year	**Number**
1961	4
1962	5
1963	3
1964	1
1965	6
1966	5
1967	1
1968	3
1969	9

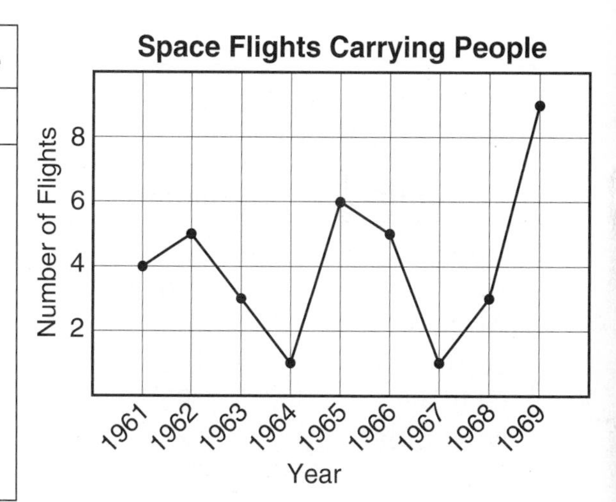

EXERCISES ***Make a line graph for each set of data.***

Sid's Daily Jogging Time for Three Miles	
Day	**Time in Minutes**
1	32
2	29
3	28
4	26
5	28
6	33
7	27

1.

Traffic on Maple Drive	
Day	**Number of Vehicles**
Monday	7,200
Tuesday	8,050
Wednesday	10,500
Thursday	5,900
Friday	9,990
Saturday	3,400
Sunday	900

Recorded Number of Hurricanes	
Month	**Number**
June	23
July	36
August	149
September	188
October	95
November	21

Evans Family Electric Bill	
Month	**Amount**
March	$129.90
April	$112.20
May	$105.00
June	$88.50

Home Runs by Hank Aaron 1967 to 1976	
Year	**Number**
1967	39
1968	29
1969	44
1970	38
1971	47
1972	34
1973	40
1974	20
1975	12
1976	10

Name ______________________ Date __________

Bar Graphs

Bar graphs are used to compare numbers. All bar graphs should have a graph title, a vertical-axis label, and a horizontal-axis label.

EXAMPLE ***Make a bar graph for the data on women's NCAA gymnastics championships between 1982 and 1993.***

NCAA Women's Gymnastics	
Year	**Champion**
1982	Utah
1983	Utah
1984	Utah
1985	Utah
1986	Utah
1987	Georgia
1988	Alabama
1989	Georgia
1990	Utah
1991	Alabama
1992	Utah
1993	Georgia

Make a tally.

Utah 𝍸 ||
Georgia |||
Alabama ||

Make a bar graph.

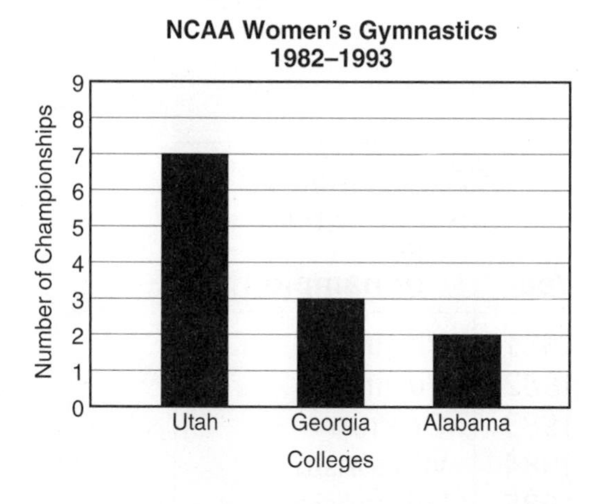

EXERCISES ***Make a bar graph for each set of data.***

1.

Preference for Brands	
Brand	**Number of Students**
A	15
B	35
C	30
D	25

APPLICATIONS

2.

NCAA Women's Volleyball	
Year	**Champion**
1981	Southern California
1982	Hawaii
1983	Hawaii
1984	UCLA
1985	Pacific
1986	Pacific
1987	Hawaii
1988	Texas
1989	California State, Long Beach
1990	UCLA
1991	UCLA
1992	Stanford

3.

NCAA Women's Cross Country	
Year	**Champion**
1981	Virginia
1982	Virginia
1983	Oregon
1984	Wisconsin
1985	Wisconsin
1986	Texas
1987	Oregon
1988	Kentucky
1989	Villanova
1990	Villanova
1991	Villanova
1992	Villanova

4. Survey the students in your math class to find out their favorite movie. Use this data to make a bar graph.

5. Survey your friends to find out their favorite television show. Use this data to make a bar graph.

Name ______________________ Date __________

Histograms

A **histogram** is a bar graph with no spaces between the bars. It shows data that has been organized into equal intervals.

EXAMPLE ***Make a histogram for the test scores on the Spanish exam.***

Test Scores on a Spanish Exam						
72	84	88	86	88	72	70
90	98	82	80	86	90	76
100	86	88	84	88	78	96
68	62	82	88	86	80	92

The scores range from 62 to 100. One possible interval that can be used to make the histogram is an interval of 10. Divide the data into the intervals 61–70, 71–80, 81–90, and 91–100.

Make a frequency chart.

Scores	Frequency
61–70	III
71–80	~~IIII~~ I
81–90	~~IIII~~ ~~IIII~~ ~~IIII~~
91–100	IIII

Draw a histogram.

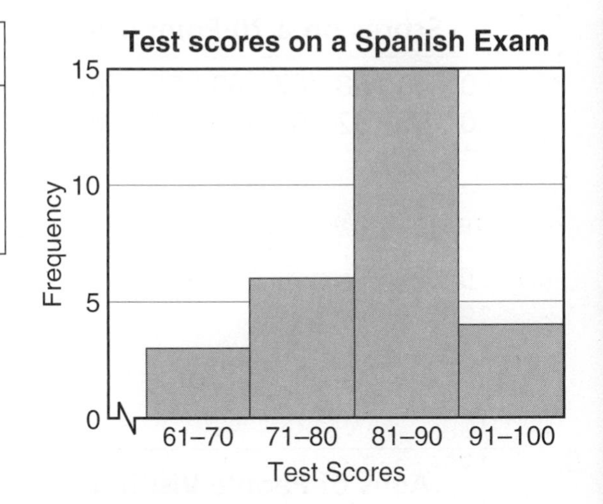

EXERCISES ***List possible intervals that could be used in making a histogram for each set of data.***

1. 782, 544, 729, 327, 489, 472, 634, 473, 379, 399, 732, 744, 799, 356, 724, 566, 532, 688, 679, 465

2. 77.3, 75.6, 76.4, 77.9, 75.8, 75.2, 76.9, 76.0, 77.3, 77.6, 76.1, 76.5, 77.5, 75.3, 75.0, 76.4, 76.2, 77.8

3. 12, 4, 6, 8, 15, 9, 2, 3, 16, 14, 7, 9, 3, 13, 14, 17, 1

APPLICATIONS ***Plan the scales and intervals for each set of data. Then make a histogram.***

4.

Height in Inches of Sells Middle School Volleyball Team							
68	69	72	64	74	56	62	58
69	65	70	59	71	67	66	64
73	78	70	52	61	68	67	66

5.

Scores on a 70-Point Science Quiz									
50	48	58	67	60	56	54	46	52	56
50	56	62	68	65	57	64	62	58	55

6.

Ages of People Visiting the Museum							
23	35	26	37	24	38	29	27
22	35	30	28	19	20	26	30
25	18	22	27	16	17	20	23

Name ______________________ Date __________

Circle Graphs

The air surrounding Earth is referred to as the atmosphere. Without air there would be no life on Earth. Air is a mixture of gases. By volume, dry air is composed of 78% nitrogen, 21% oxygen, and 1% other gases.

EXAMPLE ***Make a circle graph to show the composition of the Earth's atmosphere.***

To make a circle graph, first find the number of degrees that correspond to each percent. Use a calculator and round to the nearest degree.

Nitrogen: 78% of 360° ≈ 281°
Oxygen: 21% of 360° ≈ 76°
Other: 1% of 360° ≈ 4°

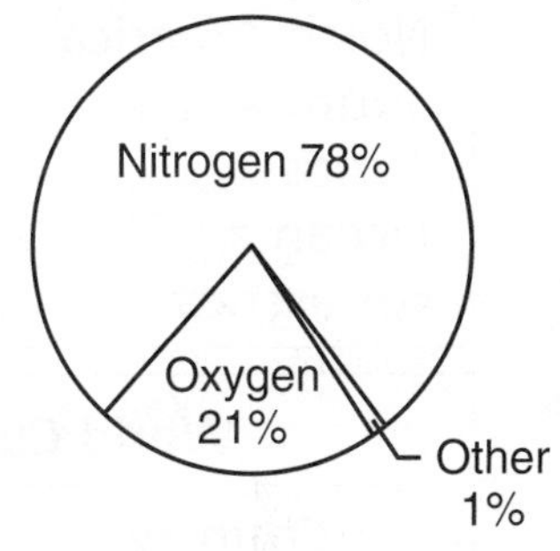

Use a compass and a protractor to draw the circle graph.

Note that the sum of the degrees is *not* 360° because of rounding.

EXERCISES ***Make a circle graph to show the data in each chart.***

1.

Favorite TV Shows	
Movies	12%
Sports	20%
News	4%
Drama	16%
Comedy	20%
Music	28%

2.

Daily Activities	
Sleeping	8 hours
Eating	1 hour
School	6 hours
Homework	3 hours
Team practice	2 hours
Miscellaneous	4 hours

APPLICATIONS ***Make a circle graph to show the data in each chart.***

3.

Area of Continents	
Continent	**Area in Millions of Square Miles**
Europe	3.8
Asia	17.4
North America	9.4
South America	6.9
Africa	11.7
Oceania	3.3
Antarctica	5.4

4.

World Cup Winners	
Country	**Number of Wins**
Argentina	2
Brazil	4
England	1
Italy	3
Uruguay	2
West Germany	3

5.

Area of New England States	
State	**Area in Square Miles**
Maine	33,215
New Hampshire	9,304
Vermont	9,609
Massachusetts	8,257
Connecticut	5,009
Rhode Island	1,214

6. Make a circle graph showing how you spent your time last Saturday.

Name ______________________ Date __________

Misleading Graphs

Graphs can be used to present data in ways that are misleading.

EXAMPLE ***A company that sells computer diskettes wants to encourage customers to buy more by showing how much the price drops as they buy more diskettes. Which of the following graphs is misleading? Which graph should the company use to encourage customers to buy more diskettes?***

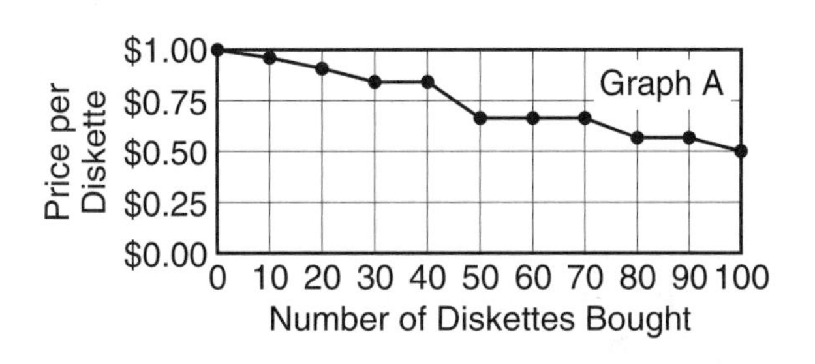

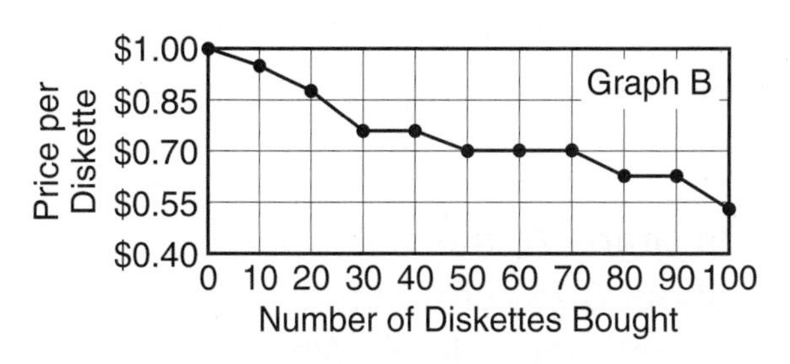

The vertical scale for Graph B does *not* begin with zero. Therefore, the drop in the cost of the diskettes seems to be greater than the actual drop as shown in Graph A. Graph B is misleading.

Since the drop in the cost of the diskettes seems greater in Graph B, this graph is more likely to encourage customers to buy more diskettes.

EXERCISES ***Use the graph at the right to answer Exercises 1–3.***

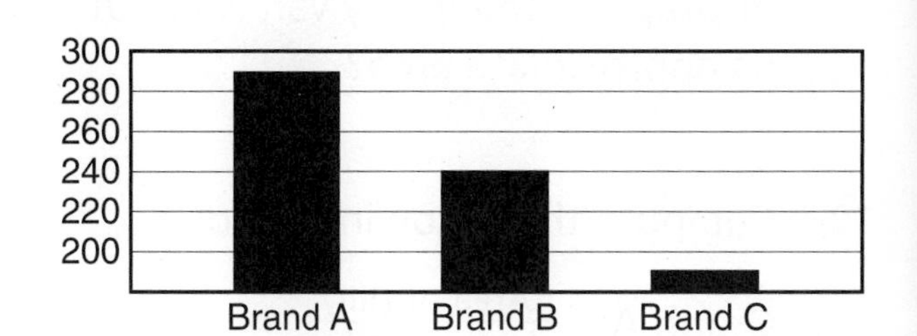

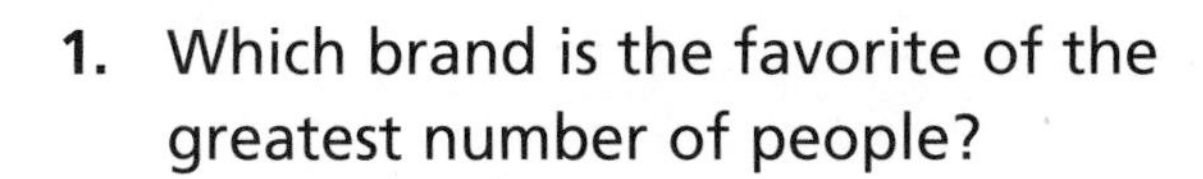

1. Which brand is the favorite of the greatest number of people?

2. Which brand is the favorite of the least number of people?

3. Why is this graph misleading?

APPLICATIONS *Use the graphs at the right to answer Exercises 4–11.*

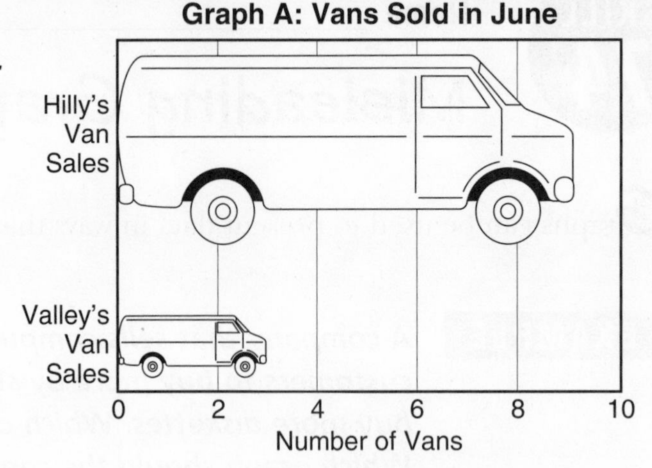

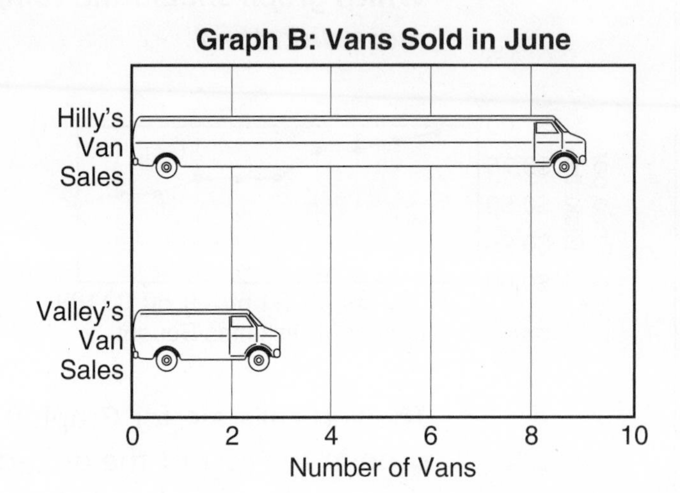

4. Do Graphs A and B give the same information on sales?

5. Find the ratio of Hilly's sales to Valley's sales.

6. In Graph A, the Hilly van is about 2.5 centimeters high by 6 centimeters long. What is its approximate area?

7. In Graph A, the Valley van is about 0.75 centimeters high and 2 centimeters long. What is its approximate area?

8. In Graph B, both vans are about 0.75 centimeter high. The Hilly van is about 6 centimeters long. What is its approximate area?

9. In Graph B, the Valley van is about 2 centimeters long. What is its approximate area?

10. Compute the following ratios.

Graph A: $\frac{\text{Area of Hilly}}{\text{Area of Valley}}$

Graph B: $\frac{\text{Area of Hilly}}{\text{Area of Valley}}$

11. Compare the results of Exercises 5 and 10. Which graph is misleading? Explain your answer.

Name ______________________________ Date __________

Tree Diagrams

The breakfast special at Dion's Place is a choice of cereal, eggs, or French toast with a choice of milk or juice for $1.99.

EXAMPLE ***If someone wishes to order a breakfast special at Dion's Place, how many choices does he or she have?***

To answer this question, make a tree diagram.

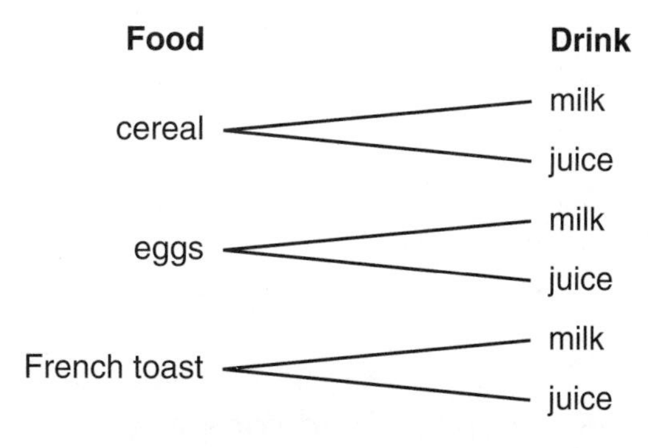

There are 6 choices for the breakfast special.

EXERCISES ***For each situation, draw a tree diagram to show all the outcomes.***

1. A number cube is rolled and a coin is tossed. What is the number of possible outcomes?

2. A penny, a nickel, and a dime are tossed. What is the number of possible outcomes?

APPLICATIONS

3. Ernie can order a small, medium, or large pizza with thick or thin crust. How many possible ways can he order the pizza?

4. Tina has a choice of a sports jersey in blue, white, gray, or black in sizes small, medium, or large. How many choices does she have?

5. José, Kara, and Beth are running for class president. Tony, Lou, and Fay are running for vice-president. How many different pairs of officers can be elected?

6. A snack food company makes chewy fruit shapes of lions, monkeys, elephants, and giraffes in red, green, purple,and yellow. How many different fruit shapes are made?

Name ______________________ Date __________

Counting Outcomes

Fundamental Counting Principle
If an event M can occur in m ways and it is followed by event N that can occur in n ways, then the event M followed by event N can occur in mn ways.

EXAMPLE ***Use three different methods to find the number of outcomes if a penny and a dime are tossed.***

Make a list.

penny, dime
heads, heads
heads, tails
tails, heads
tails, tails

Make a tree diagram.

Penny	Dime
Heads	Heads
	Tails
Tails	Heads
	Tails

Use the Fundamental Counting Principle.

outcomes for penny × outcomes for dime = possible outcomes
$2 \times 2 = 4$

There are 4 possible outcomes if a penny and a dime are tossed.

EXERCISES ***Make a list to find the number of outcomes for each situation.***

1. A coin and a number cube are tossed.

2. This spinner is spun twice.

Draw a tree diagram to find the number of outcomes for each situation.

3. Three coins are tossed.

4. A coin is tossed and the spinner in Exercise 2 is spun.

Use the Fundamental Counting Principle to find the number of outcomes for each situation.

5. Shirts come in 4 colors and 3 sizes.

6. Donna has a choice of 6 entrees and 4 beverages.

APPLICATIONS

7. The nursery has 14 different-colored tulip bulbs. Each color comes in dwarf, average, or giant size. How many possible selections are there?

8. The type of bicycle Elena wants comes in 12 different colors with 12 different colors of trim. There is also a choice of curved or straight handle bars. How many possible selections are there?

9. At a banquet, guests were given a choice of 4 entrees, 3 vegetables, soup or salad, 4 beverages, and 4 desserts. How many different selections were possible?

10. Ms. Nitobe is setting the combination lock on her briefcase. If she can choose any digit 0-9 for each of the 6 digits in the combination, how many possible combinations are there?

Name ______________________________ Date __________

Probability

The **probability** of an event is the ratio of the number of ways an event can occur to the number of possible outcomes.

$$\text{Probability of an event} = \frac{\text{number of ways the event can occur}}{\text{number of possible outcomes}}$$

EXAMPLE ***On the spinner below, there are ten equally likely outcomes. Find the probability of spinning a number less than 5.***

Numbers less than 5 are 1, 2, 3 and 4.
There are 10 possible outcomes.

Probability of number less than 5 = $\frac{4}{10}$ or $\frac{2}{5}$.

The probability of spinning a number less than 5 is $\frac{2}{5}$.

EXERCISES ***A box of crayons contains 3 shades of red, 5 shades of blue, and 2 shades of green. If a child chooses a crayon at random, find the probability of choosing each of the following.***

1. a green crayon

2. a red crayon

3. a blue crayon

4. a crayon that is *not* red

5. a red or blue crayon

6. a red or green crayon

A card is chosen at random from a deck of 52 cards. Find the probability of choosing each of the following.

7. a red card

8. the jack of diamonds

9. an ace

10. a black 10

11. a heart

12. *not* a club

A cooler contains 2 cans of grape juice, 3 cans of grapefruit juice, and 7 cans of orange juice. If a person chooses a can of juice at random, find the probability of choosing each of the following.

13. grapefruit juice

14. orange juice

15. grape juice

16. orange or grape juice

17. *not* orange juice

18. *not* grape juice

APPLICATIONS

Businesses use statistical surveys to predict customers' future buying habits. A department store surveyed 200 customers on a Saturday in December to find out how much each customer spent on their visit to the store. Use the results at the right to answer Exercises 19–21.

Amount Spent	Number of Customers
Less than $2	14
$2–$4.99	36
$5–$9.99	42
$10–$19.99	32
$20–$49.99	32
$50–$99.99	22
$100 or more	22

19. What is the probability that a customer will spend less than $2.00?

20. What is the probability that a customer will spend less than $10.00?

21. What is the probability that a customer will spend between $20.00 and $100.00?

Name ______________________ Date __________

Odds

The odds for an event can be found using the following ratio.

$$\text{odds for an event} = \frac{\text{number of ways the event can occur}}{\text{number of ways the event cannot occur}}$$

EXAMPLE ***Carol picks a marble out of a bag containing 2 red marbles, 3 blue marbles, and 5 white marbles. What are the odds that she will choose a blue marble? What are the odds against choosing a blue marble?***

The number of ways a blue marble *can* be chosen is 3. The number of ways a blue marble *cannot* be chosen is 7.

$$\text{odds of choosing a blue marble} = \frac{3}{7}$$

$$\text{odds against choosing a blue marble} = \frac{7}{3}$$

The odds of choosing a blue marble are $\frac{3}{7}$.

The odds against choosing a blue marble are $\frac{7}{3}$.

EXERCISES ***A coin is tossed. Find the odds for each of the following.***

1. tails

2. heads

3. against tails

4. against heads

A number cube is rolled. Find the odds for each of the following.

5. 1

6. 5

7. against a 1

8. against a 5

9. a prime number

10. a number less than 3

11. a number greater than 3

12. an odd number

13. not an odd number

14. a number less than 6

APPLICATIONS ***Brenda estimates that the probability that she will pass her next test is $\frac{9}{10}$. She also estimates that the probability that she will fail the course is $\frac{1}{100}$.***

15. What are the odds for Brenda passing the test?

16. What are the odds against Brenda passing the test?

17. What are the odds for Brenda passing the course?

18. What are the odds against Brenda passing the course?

19. Do you think probability or odds tell you more about how likely Brenda is to pass the test and the course? Explain.

Name ______________________________ Date ____________

Make a List

Leslie wants to take a picture of her four dogs. She has a collie, a poodle, a beagle, and a terrier.

EXAMPLE ***How many ways can she arrange her dogs in a row if the poodle and the collie cannot be next to each other?***

Let C = the collie, P = the poodle, B = the beagle, and T = the terrier. Make a list of all the possible arrangements. Then cross out any arrangement with the poodle next to the collie.

~~CPBT~~	~~PCBT~~	~~BCPT~~	~~TCPB~~
~~CPTB~~	~~PCTB~~	BCTP	TCBP
CBPT	PBCT	~~BPCT~~	~~TPCB~~
CBTP	PBTC	BPTC	TPBC
CTPB	PTCB	~~BTCP~~	~~TBCP~~
CTBP	PTBC	~~BTPC~~	~~TBPC~~

There are 12 ways the four dogs can be arranged for the picture.

EXERCISES ***Solve by making a list.***

1. How many ways can the four dogs in the example above be arranged if the beagle must always be next to the terrier?

2. How many different ways can nickels, dimes, and quarters be combined to total one dollar?

3. How many different amounts of money can you make with three dimes and three nickels?

4. List the first 20 perfect squares. Is there any pattern in the units digit?

APPLICATIONS

5. Ai-lien has a choice of chicken or spaghetti, soup or salad, and fruit or yogurt for dinner. How many possible choices does she have?

6. A kite comes in red, blue, or yellow. Strings for the kite are available in lengths of 100 feet, 120 feet, or 150 feet. How many kite/string choices can be made?

7. Yoskie, Marcie, and Shane are planning to sit in the front row of the math class. There are four seats in the front row. In how many different ways can the students seat themselves?

8. The Leshnock Dancing Center offers a choice of ballroom, tap, jazz, and ballet classes at three different levels: beginner, intermediate, and advanced. How many combinations of classes and levels are available?

9. Seth, Alfred, Miwa, Jenny, Bob, and Rachel are planning a spring party. If three students are chosen to be on the refreshment committee, how many different refreshment committees can be chosen?

10. A penny, nickel, dime, quarter, and half-dollar are in a purse. Without looking, Maria picks two coins. How many different amounts of money could she choose?

Name ______________________________ Date __________

Permutations

An arrangement or listing in which order is important is called a **permutation**.

$P(n, r)$ means the number of permutations of n things taken r at a time.

EXAMPLE ***There are 6 sailboats in a race. How many arrangements of first, second, and third place are possible?***

There are 6 choices for first place, then 5 choices for second place, and finally 4 choices for third place.

$$6 \times 5 \times 4 = 120$$

The number of permutations is 120.

Some arrangements involve all of the members of a group.

EXAMPLE ***There are 6 sailboats in a race. In how many ways can they finish the race?***

There are 6 choices for first, 5 choices for second, and so on.

$$P(6, 6) = 6 \times 5 \times 4 \times 3 \times 2 \times 1 = 720$$

There are 720 ways in which the sailboats can finish the race.

The expression $6 \times 5 \times 4 \times 3 \times 2 \times 1$ can be written 6!. It is read *six factorial.* In general, $n!$ is the product of whole numbers starting at n and counting backward to 1.

EXERCISES ***Find the value of each expression.***

1. 4! **2.** 5! **3.** $P(5, 2)$ **4.** $P(8, 3)$

5. $P(4, 3)$ 6. $P(7, 4)$ 7. $P(10, 2)$ 8. $P(4, 4)$

APPLICATIONS

9. How many ways can a winner and a runner-up be chosen from 8 show dogs at a dog show?

10. In how many ways can 5 horses in a race cross the finish line?

11. In how many ways can a president, vice-president, secretary, and treasurer be chosen from a club with 12 members?

12. A shelf has a history book, a novel, a biography, a dictionary, a cookbook, and a home-repair book. In how many ways can 4 of these books be rearranged on another shelf?

13. In how many ways can 8 people be seated at a counter that has 8 stools in a row?

14. Eight trained parrots fly onto the stage but find there are only 5 perches. How many different ways can the parrots land on the perches if only one parrot fits on each perch?

15. Seven students are running for class president. In how many different orders can the candidates make their campaign speeches?

16. In how many different ways can a coach name the first three batters in a nine-batter softball lineup?

17. How many different flags consisting of 4 different-colored vertical stripes can be made from blue, green, red, black, and white?

18. In how many ways can the gold, silver, and bronze medals be awarded to 10 swimmers?

SKILL 81

Name ______________________ Date ________

Combinations

Arrangements or listings in which order is *not* important are called **combinations**.

$C(n, r)$ stands for the number of combinations of n things taken r at a time.

$$C(n, r) = \frac{P(n, r)}{r!}$$

EXAMPLE ***In how many ways can 3 representatives be chosen from a group of 11 people?***

$$C(11, 3) = \frac{P(11, 3)}{3!}$$

$$= \frac{11 \times 10 \times 9}{3 \times 2 \times 1}$$ *Find P(11, 3).*

$$= \frac{990}{6}$$ *Divide by 3! to eliminate combinations that are the same except for order.*

$$= 165$$

There are 165 ways that 3 representatives can be chosen.

EXERCISES ***Find the value of each expression.***

1. $\frac{6!}{3!}$ **2.** $\frac{5!}{4!}$ **3.** $\frac{6!}{2!}$ **4.** $\frac{8!}{5!}$

5. $C(10, 9)$ **6.** $C(8, 8)$ **7.** $C(4, 2)$ **8.** $C(5, 3)$

9. $C(6, 2)$ **10.** $C(6, 4)$ **11.** $C(7, 3)$ **12.** $C(8, 6)$

13. $C(12, 8)$ **14.** $C(15, 6)$ **15.** $C(10, 4)$ **16.** $C(9, 3)$

APPLICATIONS

17. For an English exam, students are asked to write essays on 4 topics from a list of 8 topics. How many different combinations are possible?

18. In how many ways can a starting team of 5 players be chosen from a team of 14 players?

19. The Supreme Court has 9 justices. At least 5 of the justices must agree on a decision.

a. How many different combinations of 5 justices can agree on a decision?

b. How many different combinations of 6 justices can agree on a decision?

c. How many different combinations of 7 justices can agree on a decision?

d. How many different combinations of 8 justices can agree on a decision?

e. How many different combinations of 9 justices can agree on a decision?

f. What is the total number of ways the 9 justices on the Supreme Court can render a decision?

20. In the Illinois State Lottery, balls are numbered from 1 to 54 and put into a machine, which scrambles them. Six balls are then selected in any order. How many different ways can the winning number be chosen?

21. How many different five-card hands are possible using a 52-card deck?

Name ______________________________ Date __________

Classify Information

Some problems may contain too much information. Other problems may not contain the information you need to solve them.

EXAMPLE

Hartford, New Britain, Middletown, and Bristol form a large metropolitan area in Connecticut. In 1990, the population of Hartford was 767,841, the population of New Britain was 148,188, the population of Middletown was 90,320, and the population of Bristol was 79,488. What was the combined population of Middletown and Bristol in 1990?

What is the question?

What was the combined population of Middletown and Bristol in 1990?

What information is needed?

The populations of Middletown and Bristol in 1990 are needed.

What information is *not* needed?

The populations of Hartford and New Britain are *not* needed.

Solve the problem.

$$\begin{array}{r} 90{,}320 \\ +\ 79{,}488 \\ \hline 169{,}808 \end{array}$$

In 1990, the combined population of Middletown and Bristol was 169,808.

EXERCISES

Solve, if possible. Classify information in each problem by writing "not enough information" or "too much information."

1. If the difference of 135 and 98 is 37, what is the sum of the numbers?

2. If the sum of two numbers is 35, what is their product?

3. Find the sum of 57, 84, and another number.

4. If the product of a number and 10 is 150 and the sum of the number and 10 is 25, what is the number?

APPLICATIONS

5. Walt has 84 stamps to share with his friends. How many should he give each one?

6. Movie tickets cost $5.00 at night and $3.50 in the afternoon. Popcorn costs $2.75 and a fruit drink costs $1.75. How much does it cost to see a movie at night and buy popcorn and a fruit drink?

7. Dana's father saved $3,496 for a down payment for a new car. He bought a 6-cylinder, 4-door sedan with power steering, air conditioning, cruise control, and a stereo cassette player. The car costs an additional $9,379. What is the cost of the car?

8. There are 250 different kinds of sharks. The smallest is the black and white shark, which grows to only 6 inches long. The largest is the whale shark, which can grow to more than 40 feet long. The mako shark can swim up to speeds of 40 miles per hour. How much faster is the mako shark than the black and white shark?

9. Yosemite National Park is 759,000 acres. Zion National Park is 143,000 acres. Kings Canyon National Park is 462,000 acres. How much larger is Yosemite National Park than Kings Canyon National Park?

10. One kind of greeting card costs $2.50, while a second kind costs $1.00. If Phil bought 5 greeting cards, how much did he spend?

Name ______________________________ Date __________

Determine Reasonable Answers

Nestor has a 48-meter by 61-meter plot of land in which he wants to plant grass. He needs about one pound of seed for each 100 square meters.

EXAMPLE ***Should Nestor buy 3 or 30 pounds of seed?***

To find the amount of seed Nestor needs to buy, first estimate the area of the plot of land. The area of a rectangle is found by multiplying the length by the width.

$$\begin{array}{rcr} 48 & \text{rounds to} & 50 \\ \underline{\times\ 61} & \text{rounds to} & \underline{\times\ 60} \\ & & 3{,}000 \end{array}$$

The area is about 3,000 square meters. Divide 3,000 by 100 to find the approximate number of pounds of seed needed for the plot of land.

$$3{,}000 \div 100 = 30$$

Nestor should buy 30 pounds of seed.

EXERCISES ***Determine whether the answers shown are reasonable.***

1. $45 + 76 = 121$ **2.** $73 - 19 = 44$ **3.** $18 \times 33 = 494$

4. $972 \div 27 = 46$ **5.** $475 + 856 = 1{,}031$ **6.** $782 - 686 = 96$

7. $204 \times 57 = 11{,}628$ **8.** $3{,}708 \div 36 = 83$ **9.** $946 + 789 = 1{,}735$

10. $1,030 - 789 = 341$ 11. $77 \times 499 = 38,423$ 12. $767 \div 13 = 59$

13. $879 + 65 = 944$ 14. $807 - 455 = 452$ 15. $904 \times 66 = 49,664$

APPLICATIONS ***Solve by determining reasonable answers.***

16. There are 25 paper plates in a package. If 160 students are expected to attend a picnic, should the picnic committee buy 7 or 9 packages of plates?

17. The 20 members of the drama club are taking a trip to see a play. The cost of the trip is $450. They want to share the cost equally. Should each member contribute $20 or $23?

18. How many audio cassettes at $8.88 each can Ken expect to buy with a $50 bill?

19. Pauline buys 6 boxes of tissues containing 75 tissues each, and Mike buys 2 boxes containing 175 tissues each. Pauline guesses that she has about twice as many tissues as Mike. Is her guess reasonable?

20. A telephone call costs $0.40 for the first minute and $0.31 for each additional minute. Is $5.00 enough to pay for a 12-minute call?

21. The Parker family drove an average of 220 miles per day on their 2-week vacation. Did they travel about 3,000 miles or about 30,000 miles on their vacation?

22. Daisuke took $20.00 to the store to buy school supplies. He wants to buy 4 notebooks at $1.98 each, 2 pens at $0.89 each, 5 packages of notebook paper at $1.50 each, an eraser at $0.39, and 4 pencils at $0.10 each. Does he have enough money to buy all of these items?

Name ______________________ Date ________

Guess and Check

There are 33 members of the Kennedy Middle School Math Club. There are 7 more girl members than boy members.

EXAMPLE ***How many boys and girls are members of the club?***

Use the guess-and-check strategy to solve this problem. Suppose your first guess is 10 boys and 17 girls.

$10 + 17 = 27$

This guess is too low. Try 15 boys and 22 girls.

$15 + 22 = 37$

This guess is too high. Try 13 boys and 20 girls.

$13 + 20 = 33$

The club has 13 boys and 20 girls.

EXERCISES ***Solve by using the guess-and-check strategy.***

1. A number plus half the number is 33. Find the number.

2. What is the only number you can multiply by itself and get a product of 1,296?

3. Fill in the boxes at the right with the digits 2, 3, 4, 5, 6, and 8 to make this multiplication work. Use each digit exactly once.

$$\begin{array}{r} \square\,\square \\ \times\quad \square \\ \hline \square\,\square\,\square \end{array}$$

4. The length of a rectangle is 4 more meters than the width. The perimeter is 40 meters. Find the length.

5. The sum of two numbers is 56. The difference is 22. What are the two numbers?

APPLICATIONS

6. In the 1992 Summer Olympic Games, the Unified Team, the United States, and Germany won 115 gold medals. The Unified Team won 45 gold medals, and the United States won 4 more gold medals than Germany. How many gold medals did the United States win? How many did Germany win?

7. The Sanchez family bought tickets to the Science Museum. Admission is $8 for adults and $5 for children under 12. They spent $49 for admission. How many adult tickets and how many student tickets did the Sanchez family buy?

8. Zachary and Aimee are in a teen bowling league. The total of their bowling averages is 172. Zachary's average is 14 points higher than Aimee's average. What is Zachary's bowling average?

9. Andy has $2.80 worth of quarters and dimes in his pocket. If the number of quarters equals the number of dimes, how many quarters does he have?

10. Mei-yu bought some pens for $0.89 each and some pencils for $0.19 each. She spent $5.02. How many pens and how many pencils did she buy?

Name ______________________________ Date ____________

Draw a Diagram

The streets in Sachi's city are arranged in square blocks. Sachi left her house and walked 5 blocks east and 2 blocks north to Jim's house. She and Jim walked 1 block north, 3 blocks west and 1 block north to school. After school, Sachi walked 2 blocks west and 2 blocks south to Bella's house.

EXAMPLE ***How far from Bella's house is Sachi's house?***

Draw a diagram to show Bella's route.

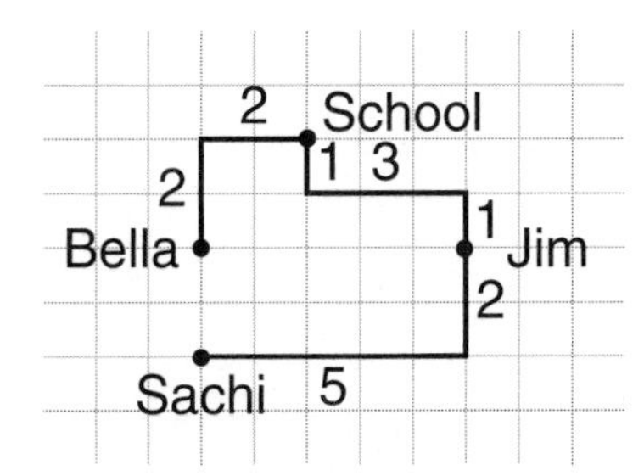

From the diagram, Bella's house is 2 blocks north of Sachi's house.

EXERCISES ***Solve by drawing a diagram.***

1. The streets in Brad's town are arranged in square blocks. Brad leaves his house and rides his bicycle 4 blocks south and 5 blocks west to the athletic field. He then rides 2 blocks south and 5 blocks east to the library. How far is the library from Brad's home?

2. How many different rectangles can be made with 36 identical square tiles if you use all the tiles?

3. Five points are marked around a circle. How many straight lines must you draw to connect every point with every other point?

4. An octagon is a polygon with 8 sides. How many diagonals does an octagon have?

APPLICATIONS

Use the menu below to answer Exercises 5–7.

WENDY'S DINER	
Sandwiches	**Drinks**
ham	lemonade
turkey	iced tea
tuna	milk
peanut butter	**Desserts**
corned beef	cookie
	ice cream

5. Sam plans to order a sandwich and a drink. How many choices does he have?

6. Rosa plans to order a sandwich and a dessert. How many choices does she have?

7. Enrique plans to order a sandwich, a drink, and a dessert. How many choices does he have?

8. Coach Williams wants to schedule a round-robin tournament (every student plays every other student) for his chess club. If there are seven club members, how many games should the coach schedule?

9. After a little-league baseball game, each player must shake hands with every player on the opposing team. If each team has 12 members, how many handshakes are there?

10. Nora arranged five friends in a line for a photograph. Bill stood between Ellen and Tomas. Sam stood between Tomas and Clara. Clara was on the left. In what order were the five people arranged for the photograph?

Name ______________________ Date ____________

Eliminating Possibilities

Ben wants a part-time job. There are three 10 hour per week jobs available. One pays \$45 per week. One pays \$4.75 per hour. One pays \$3.25 per hour plus tips which average \$20 per week.

EXAMPLE ***If Ben wants to take the job that pays the most, for which job should he apply?***

To solve this problem, find the weekly pay for each job. Then eliminate the lowest paying jobs.

\$4.75 per hour × 10 hours = \$47.50 per week.
That is more than \$45 per week. Eliminate the job that pays \$45 per week.

\$3.25 per hour × 10 hours = \$32.50
\$32.50 + \$20 = \$52.50 per week
That is more than \$47.50 per week. Eliminate the job that pays \$47.50 per week.

Ben should apply for the job that pays \$3.25 per hour plus tips.

EXERCISES ***Solve by eliminating possibilities.***

1. Jon bought 15 dozen pencils. This number of pencils is about:

a. 45 **b.** 125 **c.** 175 **d.** 1,000

2. There is about \$1.66 in one pound of pennies. Kent has 16 pounds of pennies. Kent has about:

a. \$25 **b.** \$16 **c.** \$165 **d.** \$50

3. The odometer in Timothy's automobile registered 68,364.7 miles last month. The odometer registers 71,429.2 miles today. Last month, the number of miles the automobile traveled was about:

a. 900 b. 3,000 c. 10,000 d. 71,000

APPLICATIONS

4. Jackie used her calculator to multiply 678×34. Should she expect the product to be about 2,305, 23,050, or 230,500?

5. Micaela gave the clerk $50 to pay for a $21.79 hat and a $17.33 belt. Should she expect $10, $20, or $40 change?

6. If Carl blinks 15 times per minute, does he blink about 216, 2,160, or 21,600 times per day?

7. Juan bought a midsize car with a $1,000 down payment and a plan to pay $210.89 per month for 48 months. Is the total cost of the car closer to $5,000, $8,000, $11,000, or $14,000?

8. Joanna's quiz scores were 23, 14, 18, 15, 21, and 20. Is the best estimate of her average quiz score 15, 18, 20, or 22?

9. A circle has a radius of 6 feet. Is the circumference closer to 40 feet, 70 feet, 90 feet, or 120 feet?

Name ______________________ Date __________

Using a Matrix

Connie, Kristine, and Roberta are the pitcher, catcher, and shortstop for a softball team, but not necessarily in that order. Kristine is not the catcher. Kristine and Roberta share a locker with the shortstop.

EXAMPLE ***Who plays each position?***

Write *no* in Kristine's row under catcher. Write *no* in Kristine's row and in Roberta's row under shortstop. Use logic to decide who plays each position.

	Pitcher	Catcher	Shortstop
Connie	no	no	yes
Kristine	yes	no	no
Roberta	no	yes	no

Connie is the shortstop, Kristine is the pitcher, and Roberta is the catcher.

EXERCISES ***Solve each problem.***

1. Ana, Iris, and Oki each have a pet. The pets are a fish, a cat, and a bird. Ana is allergic to cats. Oki's pet has 2 legs. Whose pet is the fish?

2. A banker, a cook, and a farmer are named Bill, Carl, and Fred, but not necessarily in that order. No one's job starts with the same letter as his name. Bill bought eggs from the farmer. Who is the banker?

3. Dan, Nan, and Fran have lockers next to each other. Nan rides the bus with the person whose locker is at the right. Dan's locker is not next to Nan's locker. Who has the locker at the left?

4. Yaizo, Kristin, and Brian each play one of the following sports: tennis, basketball, and swimming. Brian and the basketball player live on the same street. Kristin is the tennis player. Who is the swimmer?

APPLICATIONS

5. Luis Alvarez, Franklin Chang-Diaz, Margarita Colmenares, Antonia Novello, and Severo Ochoa have made important contributions to science and engineering. One is an astronaut, one a biochemist, one an engineer, one a physician, and one a physicist.

   Alvarez was the winner of the Nobel Prize for Physics in 1968.

   Ochoa won the Nobel Prize for Physiology or Medicine in 1959, but he is not the physician.

   The engineer is the first woman president of the Society of Hispanic Professional Engineers.

   The physician is the first woman surgeon general of the United States.

   Colmenares is not the physician.

   Ochoa was born 45 years before the astronaut.

   Who is the astronaut?

Name ______________________ Date __________

Look for a Pattern

Sergio starts a savings account by depositing \$1.00 into his account the first week, \$3.00 the second week, \$5.00 the third week, \$7.00 the fourth week, and so on.

EXAMPLE ***How much money will he have in his savings account on the twentieth week?***

Find the total amount Sergio will have in his account on each of the first four weeks.

Week	1	2	3	4
Deposit	\$1	\$3	\$5	\$7
Total	\$1	\$4	\$9	\$16

Look for a pattern. The totals are squares of the numbers of the weeks.

$1^2 = 1 \qquad 2^2 = 4 \qquad 3^2 = 9 \qquad 4^2 = 16$

Use the pattern to find the total on the twentieth week.

$$20^2 = 400$$

Sergio will have \$400 in his savings account on the twentieth week.

EXERCISES ***Write the next two numbers in each pattern.***

1. 17, 34, 51, 68, . . .

2. 3, 6, 12, 24, . . .

3. 113, 106, 99, 92, . . .

4. 20, 22, 25, 29, 34, . . .

5. $\frac{1}{2}, \frac{1}{3}, \frac{1}{4}, \frac{1}{5}, \ldots$

6. 8, 0, 8, 0, . . .

Solve by finding a pattern.

7. What is the total number of rectangles in the figure at the right?

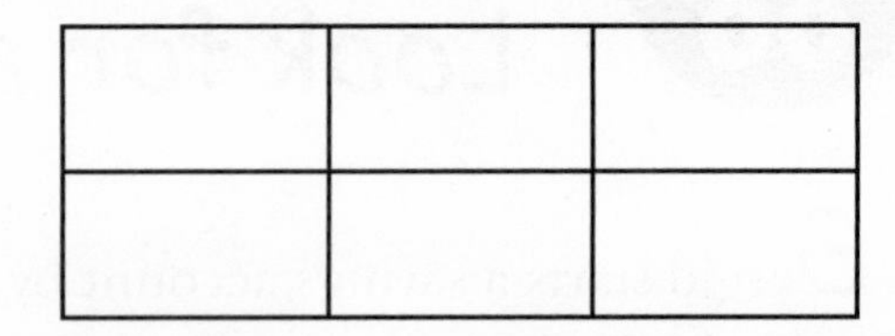

8. How many diagonals does a ten-sided polygon have?

APPLICATIONS

9. Isabel wants to work up to doing 40 sit-ups a day. She plans to do 7 sit-ups the first day, 10 the second day, 13 the third day, and so on. On what day will she do 40 sit-ups?

10. The NCAA men's basketball tournament starts with 64 teams. After the first round, there are 32 teams left; after the second round there are 16 teams left, and so on. Complete the pattern until there is only one team left. How many rounds does it take to determine a winner?

11. Alfua has a starting salary of $21,500. She receives annual raises equal to $\frac{1}{10}$ of her current salary. How many years must she work to double her salary?

12. The bus leaves downtown for the mall at 7:35 A.M., 8:10 A.M., 8:45 A.M., and 9:20 A.M. If the bus continues to run on this schedule, what time does the bus leave between 10:00 A.M. and 11:00 A.M.?

13. Choose three different digits. Use these digits to make all possible two-digit numbers in which the tens digits and the ones digit are different (six different numbers). Add them. Add the three original digits. Divide the first sum by the second sum. What is the answer? Repeat this procedure three more times, each time using a different group of three digits. What is the pattern in the answers?